INTEGRATED ENVIRONMENTAL MANAGEMENT

Development, Information, and Education in the Asian-Pacific Region

INTEGRATED ENVIRONMENTAL MANAGEMENT

Development, Information, and Education in the Asian-Pacific Region

GE
90
.P33
I58
1999
west

Edited by
Yasumasa Itakura
J.S. Eades
Frank M. D`Itri
Munetsugu Kawashima
Shuichi Endoh
Hiroaki Kitamura

LEWIS PUBLISHERS
Boca Raton London New York Washington, D.C.

Library of Congress Cataloging-in-Publication Data

Catalog record is available from the Library of Congress.

This book contains information obtained from authentic and highly regarded sources. Reprinted material is quoted with permission, and sources are indicated. A wide variety of references are listed. Reasonable efforts have been made to publish reliable data and information, but the author and the publisher cannot assume responsibility for the validity of all materials or for the consequences of their use.

Neither this book nor any part may be reproduced or transmitted in any form or by any means, electronic or mechanical, including photocopying, microfilming, and recording, or by any information storage or retrieval system, without prior permission in writing from the publisher.

All rights reserved. Authorization to photocopy items for internal or personal use, or the personal or internal use of specific clients, may be granted by CRC Press LLC, provided that $.50 per page photocopied is paid directly to Copyright Clearance Center, 222 Rosewood Drive, Danvers, MA 01923 USA. The fee code for users of the Transactional Reporting Service is ISBN 1-56670-419-7/99/$0.00+$.50. The fee is subject to change without notice. For organizations that have been granted a photocopy license by the CCC, a separate system of payment has been arranged.

The consent of CRC Press LLC does not extend to copying for general distribution, for promotion, for creating new works, or for resale. Specific permission must be obtained from CRC Press LLC for such copying.

Direct all inquiries to CRC Press LLC, 2000 N.W. Corporate Blvd., Boca Raton, Florida 33431.

Trademark Notice: Product or corporate names may be trademarks or registered trademarks, and are used only for identification and explanation, without intent to infringe.

© 1999 by Shiga University
Lewis Publishers is an imprint of CRC Press LLC

No claim to original U.S. Government works
International Standard Book Number 1-56670-419-7
Printed in the United States of America 1 2 3 4 5 6 7 8 9 0
Printed on acid-free paper

FOREWORD

In the 21st Century mankind will have to solve many environmental problems. For this purpose efforts by many people will be necessary, one of which is to promote environmental education. Based on the experience of environmental education, it is preferable that the practice and subject matter be discussed along with the role of information systems, distance education, and environmental information networks, and how these systems can be delivered globally.

The 1997 Shiga University International Symposium had the theme, "What can education and information systems do to solve environmental problems?" The objectives are summarized in the following three points:
1) Identify the Asian-Pacific region's environmental problems and develop programs to promote changes in environmental awareness
2) Review environmental education experiences and identify problems still unsolved
3) Define the role of information systems, distance education, and environmental networks in achieving these goals/objectives.

This book is the outgrowth of an international symposium which was convened at the Otsu City Lifelong Learning Center in Otsu and the Japan Center for Michigan Universities in Hikone, Shiga Prefecture, Japan, November 22-24, 1997. The papers can be divided into three groups which discuss, in turn, environmental problems and environmental awareness; experiences of environmental education; and the information systems related to these. The final paper addresses perspectives on these themes and provides a general discussion. It is hoped that this book will assist in defining the nature of the problems and contribute to their solution in the next century.

Still, much remains to be learned about the promotion of environmental education and supporting systems of information technology. Therefore, we intend to convene similar symposia to study and discuss them in future. Through these efforts Shiga University may be able to contribute more fully to the international community.

Shiga University has cooperation and exchange agreements with Michigan State University, Deakin University, the University of Glasgow, and Berlin Technical and Economics University, and it has made efforts to promote international collaboration and student exchanges with these universities. Colleagues from these universities have previously been engaged in joint research projects with us relating to environmental sciences, environmental education, and distance learning. This background provided the necessary basis for holding this symposium. Additionally, it was held under the auspices of Shiga Prefecture Government, Shiga Prefectural Board of Education, Otsu City Government, Otsu Municipal Board of Education, Hikone City Government, Hikone Municipal Board of Education, and the Environment Agency of Japan. I express our sincere thanks for their support.

Mikita Kato
President, Shiga University
October, 1998

PREFACE

The 1997 Shiga University International Symposium

In the 21st Century, the human race will face what the Club of Rome (1972) has termed "the Great Transition". The problems involved in this transition include explosive population growth, destruction of the natural environment, and the precarious nature of the food supply, to name but a few. These are global problems, which need to be attacked on a broad front if they are to be solved. A combination of natural, human, and social resources is necessary in attacking them.

Shiga University is in Shiga Prefecture, which has the largest lake in Japan, Lake Biwa. This is one of the world's most important natural resources, surpassing all other lakes in terms of the range and number of human needs which it meets. The management of Lake Biwa presents severe problems with regard to resource development and environmental preservation which have to be resolved. Shiga University has, therefore, concentrated its research on environmental problems and education seen from a broad perspective. One of our efforts is to promote international collaboration. For example, Professor Norio Suzuki and his colleagues were engaged in a joint research project on wetland environments with Professor Frank M. D'Itri who is Associate Director of the Institute of Water Research at Michigan State University, USA, and they have obtained interesting results. At present, Professor Munetsugu Kawashima and his colleagues are conducting joint research on environmental education with researchers at Chiang Mai University, Thailand. Professor Yasumasa Itakura and his colleagues are also conducting joint research on distance education with researchers at Deakin University, Australia. These research projects have been carried out with a long-term research grant for joint international research with other universities from the Japanese Ministry of Education.

These research projects were initially conducted independently from each other so it was also necessary to provide opportunities to exchange ideas and research findings more broadly. We needed to publicize the results of this research and establish new areas for international collaboration. Accordingly, Shiga University planned and held an international symposium to discuss environmental awareness, education, and distance learning. The theme of this symposium was the ways in which education and information systems can contribute to solving environmental problems. The practice and subject matter of environmental education were discussed, along with the role of information systems, distance learning, and environmental networks, and how these systems can be promoted globally.

The coordinators of the symposium organized the discussion of this theme under three headings. The first is that of the Asian-Pacific region's environmental problems and how to induce changes in environmental awareness. Each country or area in the Asian-Pacific region has its own particular environmental problems, which need to be understood, together with changing patterns of environmental awareness. Based on these considerations, the symposium investigated the tasks which education has to undertake in order to find appropriate solutions.

The second is that of the experience of environmental education in the Asian-Pacific region and the related problems which still need to be solved. The main purposes of environmental education are to promote understanding of the present situation and its problems and to cultivate the practical ability of the people to solve them. It is necessary to discuss ways of improving educational materials, methods, technology, and teaching and learning systems, so as to promote environmental education.

The third is that of the role of information systems, distance learning, and environmental networks in supporting environmental education worldwide. We must "think globally and act locally" in relation to environmental problems, but we must also discuss them with the rest of the world. Not only do we have the opportunity to exchange ideas and results as citizens at the local level, but we also have to discuss the problems globally to promote international cooperation to find appropriate solutions. Distance learning and the environmental networks supported by information technology can play an important role in accomplishing this task.

The symposium had in all seventeen invited papers and about 200 participants, including twelve from other countries. It was successful in presenting valuable summaries of the region's environmental problems and suggestions for solving them, and these are presented in this book. The twenty-one chapters include revised versions of all the papers presented at the symposium. We hope that through this book Shiga University can help shed light on these subjects and contribute to the work of the international community.

On behalf of the editorial committee established to arrange publication at the time of the symposium, we wish to thank all the authors of the papers whose research, time, effort, and advice made the symposium a success and this book possible: Masahisa Nakamura,

Lake Biwa Research Institute; Sham Sani, University Kebangsaam Malaysia; Azra Meadows and Peter S. Meadows, The University of Glasgow; Junk Wk Kim, Kye Won Lee, and Gyu Ho Jung, Seoul National University; Supachit Manopimoke, Thammasat University; Tatuo Kira, International Lake Environment Committee Foundation; Sirmsree Chaisorn, Chiang Mai University; Merle C. Tan, University of the Philippines; Shigeki Shirai, Shiga Prefectural Board of Education; Mohamad Soerjani, National Research Council, Indonesia; Tomitaro Sueishi, The University of Shiga Prefecture; Edwin Brumby, Deakin University; Takashi Sakamoto, National Institute of Multimedia Education; Daryl Le Grew, Deakin University (now at the University of Canterbury, New Zealand); Takashi Toda, Lake Biwa Museum, and Kazuyuki Matsui, Shiga Prefectural Education Center; Hajime Hirai, Shiga University; Suraphol Sudara, Chulalongkorn University.

Yasumasa Itakura
Shiga University
Otsu, Japan

Reference
Club of Rome. 1972. *The Limits to Growth, A Report for the Club of Rome's Project on the Predicament of Mankind.* Earth Island Ltd. , London, UK.

CONTENTS

1	High-Speed Growth, Politics and the Environment in East and Southeast Asia.	*J.S. Eades*	1
2	Lake Biwa and the Asian Environmental Agenda: Issues and Prospects.	*Masahisa Nakamura*	19
3	Beyond Environmental Legislation: Environmental Education in Malaysia	*Sham Sani*	29
4	Mountains, Rivers, and the Coastal Zone in Asia: Environmental Management and Community Integration into the 21st Century.	*Azra Meadows and Peter S. Meadows*	49
5	Environmental Problems and Public Awareness in the Republic of Korea.	*Jung Wk Kim, Kye Won Lee and Gyu Ho Jung*	77
6	Economic Development and the Environment in Thailand: The Current Situation and the Role of Environmental Education and NGOs.	*Supachit Manopimoke*	93
7	Problems of Ecosystem Devastation as a Focus in Environmental Education.	*Tatuo Kira*	131
8	Education for Solving Environmental Problems: How to Develop Teaching Materials and Generate Support for Environmental Education in Schools.	*Munetsugu Kawashima*	143
9	The Present Situation and Problems of Environmental Education in Shiga Prefecture.	*Shigeki Shirai*	153
10	The Development of Environmental Education Curricula and Some Examples of Thai Experiences.	*Sirmsree Chaisorn*	163

11	In-Service Training Programs in Environmental Education in the Philippines — The Role of UP-ISMED.	*Merle C. Tan*	169
12	Local Involvement through the Environmental Network: The Key to Successful Natural Conservation.	*Suraphol Sudara*	181
13	Environmental Education in Support of Sustainable Development: The Case of Indonesia.	*Mohamad Soerjani*	189
14	Literacy Regarding Environmental Issues — Allonomy or Autonomy?	*Tomitaro Sueishi*	199
15	Environmental Contamination and the Information Highway.	*Frank M. D'Itri*	211
16	Networks, Technology, Distance Learning and Environmental Education.	*Ed Brumby*	223
17	The Use of Advanced Information and Communication Technologies in Education.	*Takashi Sakamoto*	229
18	Links, not Boundaries: An Asian-Pacific Environmental Education Network.	*Daryl Le Grew*	241
19	Biwako-Das: Public Collaboration in Meteorological Observation with a Computer Communication Network.	*Takashi Toda and Kazuyuki Matsui*	253
20	Establishing a Course in the Practical Use of Media Tools at Shiga University — Introduction of a Flexible Learning Program.	*Hajime Hirai*	261
21	What Education and Information Systems Can Do to Help Solve Environmental Problems: A Summary.	*Frank M. D'Itri*	271
	Index		279

Chapter 1

High-Speed Growth, Politics and the Environment in East and Southeast Asia

J.S. Eades, Department of Social Systems, Faculty of Economics, Shiga University, Hikone, Shiga Prefecture 522-0069, Japan; Department of Anthropology, University of Kent, Canterbury CT2 7NS, UK.

Introduction

In this paper I want to review some of the literature on the environmental problems of East and Southeast Asia, focusing on two main questions:
1) What have been the consequences of high speed growth for the environments of the region?
2) What are the political processes through which these problems are being alleviated or may be alleviated in the future?

It seems self-evident that there has been severe environmental damage as a result of high-speed growth in virtually all the countries of the region, starting with Japan and extending to Korea, Taiwan, China, and more recently Malaysia, Indonesia and Thailand, as a result of either capitalist or socialist policies which put too much emphasis on economic growth, without dealing with the environmental consequences of it. This has led to a realization, beginning with Japan, and continuing with Taiwan, Korea, and even China, that attempts have to be made in the planning process to avoid possible adverse environmental consequences. There have also been attempts in some countries to accommodate or even encourage the activities of NGOs, pressure groups, citizen's movements etc. concerned with the environment for a variety of political reasons.

In this paper, therefore, I will survey some of the literature on environmental damage in the countries of the region, look at the root causes of it in local economic and political conditions, and consider the various types of political activism which it has led to. Finally, at the end of the paper I will consider the continuing obstacles to alleviation which still exist, not only in countries like China which are only just beginning to move away from policies of high-speed industrialization, but even in Japan.

Economic Development and Environmental Degradation

In the postwar period the economy of East Asia has developed very rapidly. Starting with Japan and the postwar reconstruction of the 1940s and 1950s, it continued with the period of Japanese high speed growth in the 1960s, the emergence of the four "little tigers" of Taiwan, South Korea, Singapore and Hong Kong in the 1970s and 1980s, and the rapid economic growth of China from 1978 to the present. Other countries in the region have also experienced rapid growth — or at least they did until last year — including Malaysia, Indonesia, Thailand, Vietnam and so on. Some economists have likened the economies of East Asia to a flock of flying geese, with Japan flying in front, followed by Singapore and Hong Kong, then followed by Taiwan and Korea, and now by China and the others (Overholt, 1993).

The problem is though that if you look at the places where these geese have flown, there is a lot of mess left behind them. The problems became evident quite early on. They are illustrated in a very concrete form in the changes which the American anthropologist Robert Smith saw in northern Shikoku, in between his first fieldwork in the early 1950s and a return visit a quarter of a century later (Smith, 1978: 7-11). In the 1950s the Inland Sea was an extraordinarily beautiful body of water, with splendid views of green islands and surrounding mountains. By the 1970s, land reclamation projects had taken place all along the shores, and petro-chemical plants, oil refineries and storage tanks had appeared, together with their associated shipping and pollution. Most of the marine life had died, and much of what remained was inedible. The fishing industry had disappeared. So had most of the vegetation on the islands, not that they could be seen very often because of air pollution. Of course the reason why the islands had been so green in the 1950s was the collapse of the local economy during and after the war, and the people themselves clearly preferred their more comfortable lives in the 1970s, despite the pollution and the overcrowding.

Smith's account illustrates well many of the political and economic problems surrounding economic development. As in other rapidly industrializing countries it is clear that economic development has usually involved a tradeoff between environmental quality and quality of life in terms of consumer goods and standard of living. However, there also comes a point when the environment deteriorates to such an extent that the quality of life is harmed anyway. Amidst the general pollution and concerning of the Japanese landscape in the postwar period, there took place some spectacular environmental disasters (McKean, 1982). These triggered off not only determined political action by the citizens affected, but also generally raised environmental awareness and the readiness with which people were able to mobilize to protect the environment in other parts of the countryside less badly affected. The "big four" lawsuits of the late 1960s and early 1970s resulted in verdicts in favor of the victims: of Itai-itai disease in Toyama, organic mercury poisoning at Minamata in Kumamoto and Niigata, and Yokkaichi asthma in Mie. The Japanese government passed anti-pollution legislation in 1970 in an attempt to solve the problems, but the environment remained an important issue and mass citizens' movements continued to proliferate.

McKean argues that these protests often became political protests instead of merely lawsuits against companies because of the practice by which local governments encouraged local industrial developments as ways of creating jobs and revenue. During the postwar period, the central government had increasingly devolved responsibilities over services and infrastructure to local government without the cash to pay for them, which made revenues from industry more important. Local politicians assumed that deference to central government was a good thing and they tried to avoid conflict, but when local problems of pollution etc. arose, they found themselves under attack from the local people. McKean's political message in 1982 was clear: more and more people were realizing through participation in these movements that they could struggle effectively for their rights, and this was beginning to have an effect on the support and the policies of the government.

As a result the government has been forced into changing its policies away from development, and more towards a mitigation of environmental damage and restoration of the environment to its former condition where possible. Some of these efforts have been successful: given Japan's current role in the international economy new industries currently coming on stream tend to cause less environmental damage than the heavy industries of the early postwar period. However, even now many of the development schemes currently underway have potentially negative consequences for the environment, and they are being pushed ahead despite public protest. In the final section of the paper I consider why this might be so, as well as factors which might bring an end to this in the future. For the moment however, I am concerned with the early postwar period, the developments which took place, and the gradual shift from an emphasis on economic growth to one on taking account of environmental constraints in the project planning process.

Japan's Experience in Comparative Perspective

How far has Japan's experience been followed in the other countries of the region? The papers in this volume on Korea, Malaysia, Indonesia and Thailand make clear that both the pattern of economic growth and its impact on the environment also have similarities there. In some cases there are direct historical links as well. Kim's paper (Chapter 5) shows that many of Korea's environmental problems stem from the colonial period, when most of the country's trees were cut down. Further damage was done in the postwar period with the development of heavy industry, which, as in Japan, also led to increasing pollution, especially in industrial areas and river basins with intensive industrial development. And, as in Japan, wetland reclamation and golf course construction are also common. The emission of atmospheric pollutants, and the output of industrial waste water and solid wastes have also grown rapidly. The increasing number of vehicles has resulted in smog, and the quality of drinking water is low. A further problem for Korea is the proximity of China and its environmental problems which I discuss later on. As in Japan public awareness of the problems also began to grow, despite media censorship, though political mobilization around environmental issues was impossible until the democratization process of the late 1980s. Also as in Japan, pollution incidents have continued, despite the public protests, and public consultation before development projects is not always carried out.

Sani's paper (Chapter 3) shows similar trends in Malaysia, as the country has moved from being a primary producer of rubber and tin to a diversified industrial economy. One major environmental concern has been forest depletion, and forests now cover less than half the land area, compared with over two thirds in the 1960s. The main culprits have been conversion of land to agriculture, dam construction resulting in flooding, mining, logging, and shifting cultivation. The same processes have been responsible for damaging the country's biodiversity, especially along the coast. Air pollution has risen in urban areas, thanks to motor vehicles, power stations and waste incineration. Rivers have also become increasingly polluted.

The Malaysian government has responded with an impressive list of environmental control legislation, and environmental concerns have been increasingly integrated into development plans. This is in addition to support for environmental monitoring, education, research, cooperation and project coordination by government agencies, and initiatives at the regional level.

Both the level of development and the problems are similar in Thailand, where rapid economic growth has been heavily dependent on exploitation of the country's resource base. Here again the main problem has been deforestation, with half the country's forests being cleared between 1960 and 1993. According to Manopimoke's paper (Chapter 6), much of it was converted to agricultural land, but illegal logging continues, and there are worries that the forest will disappear completely in the next generation. In order to protect these areas and their biodiversity, the government has increased the area of the national parks, but without proportionate provision of cash to maintain them.

Thailand also suffers from problems with the agricultural land, about half of which is affected by soil erosion, salinity and acidity. Much of the most fertile land has been affected by urbanization, with building for residential, commercial and industrial purposes, especially around the major cities, and pollution has also increased as a result of urban waste. Urban growth has also put pressure on the water supply, and the country's coastal problems are also similar to those of Malaysia.

As with Malaysia, earlier national development plans in Thailand tended to concentrate on economic growth, but later plans, from the mid-1970s, have recognized some of the environmental problems and have attempted a degree of protection and rehabilitation. The Sixth Plan (1981-86) marked an important turning point in that it recognized the need for non-agricultural sources of income, and the need to protect the forests from encroachment, and the Seventh and Eighth Plans (1991-96, 1997-2003) have taken this even further. However, there have been continued failures in monitoring and enforcement, making the role of education, public awareness and the role of NGOs that much more important. Informal education, through religious and community organizations has also been important as a resource for the environmental movement. Generally, however, the picture is much the same as in Malaysia: the government generally makes the plans, and the people fall in line with them. The community and religious organizations which Sudara (Chapter 12) describes may well be the exception rather than the rule.

As the largest country in the region apart from China, Indonesia suffers from very similar environmental problems to those of its neighbors. These became particularly apparent with the haze crisis of late 1997, in which smoke from the burning Indonesian forests set alight by shifting cultivators, in addition to the unusual climatic conditions following El Niño, led to the spread of air pollution in the form of thick haze which affected its neighbors Malaysia, Singapore and Thailand in addition to large parts of Indonesia itself. But Indonesia's problems are probably also accentuated because of the size of the country, the uneven nature of economic development, and the political corruption under the previous regime, which have made monitoring and law enforcement even more difficult than in Thailand or Malaysia. As Soerjani's paper (Chapter 13) makes clear, Indonesia has a clear appreciation of the nature of the problems and the necessity for environmental education to increase public awareness of environmental issues, especially those that are related to the exploitation of primary commodities such as minerals and timber. However, as he also points out, Indonesia is currently marginally unsustainable. Illegal logging, collusion between villagers, officials, and logging companies, and the flouting of environmental regulations are hastening the environmental deterioration. Finally there is an international dimension to the problem, in that much of the wood ends up as wood pulp for shipment to Japan, which has financed the pulp mills through its international development program. Indonesia's rate of forest depletion is, therefore, much higher than that of Japan! Under the Suharto regime of course, effective citizen opposition to the government's plans for the environment was generally impossible.

Finally, as the haze problem in the last couple of years has shown, Malaysia, Singapore and Thailand can also be badly affected by what goes on over the border in Indonesia. Solutions to environmental problems increasingly depend on international and not just national action.

Taiwan

As with the other countries discussed above, Taiwan has moved from indifference to the environment in the 1950s and 1960s to a concern for protecting it at the present day. The Environmental Protection Agency, established relatively recently, stresses the need for integrating protection measures as part of development planning. This adds 3% to the cost of projects, compared with clearing up the mess afterwards which increases costs by 40%.

As Williams makes clear (1994), Taiwan's environmental problems and the way it has dealt with them have been rather similar to Japan, with a laissez faire attitude until the 1970s gradually replaced by a realization of the problems which had been caused.

Even the economists who described Taiwan's postwar economic growth tended to ignore the environmental impact almost entirely until comparatively recently. His discussion of the causes of Taiwan's environmental problems can be summarized roughly as follows.

As with Japan, Taiwan is very crowded and for similar reasons: the population of 20 million is squeezed onto an island which is mainly mountainous. The parts that are flat are among the most densely populated in the world, with around 1600 people per square kilometer. In some of the wealthier countries, e.g. Holland and Belgium in Europe, the environment has been successfully managed despite similar population densities. However, in the case of Taiwan, because of the very rapid economic growth there has been insufficient monitoring of the environment, and so the problems have not been controlled, and some critics have argued that many Taiwanese are more interested in the profits of their own little family businesses than in the environmental welfare of the whole population. This may not only be a problem in Taiwan: in Indonesia and China as well, the struggle of families to survive is an important component of the environmental crisis.

The international division of labor also means that much of the industrialization in Taiwan, like that in Korea, has been of the type that causes large amounts of pollution: engineering, chemicals, etc. But in the case of Taiwan there have been two other problems. First, many of the industries are located in smaller towns and rural areas. This has had the benefit of decentralizing production and reducing the pressure on the main cities, though it has made the environmental problems of the countryside worse. Second, many of the companies are apparently small, and interested in cutting costs to compete, so they have not been very interested in raising their production costs by controlling pollution.

These developmental processes have led to a wide variety of pollution problems of which the main ones are biologically active wastes, inert and semi-inert substances, hazardous wastes, and atmospheric pollution.

In relation to biologically active wastes, there has until recently been very little treatment of sewage in Taiwan, and this has had an impact both on human health, with high rates of hepatitis, and on water resources, with eutrophication and damage to the fish in most major rivers and many reservoirs. It is clear what the problem is and what to do about it, but the big question is: who will pay for cleaning it up?

As for inert and semi-inert substances such as plastics, glass, and metal, as economic growth has continued, so has the output of these substances in garbage. Most garbage is disposed of in landfills, but these otherwise harmless materials provide ideal breeding grounds for rats, flies and other pests, which once more have an impact on health. Landfill sites in the Taipei area are in any case increasingly difficult to find. The alternatives are incineration, which is expensive and could lead to air pollution, and separation of garbage coupled with recycling, which is apparently also seen as difficult to manage in Taiwan, although it is successfully employed in Japan.

Hazardous wastes come mainly from industry and include chemicals, pesticides, radioactive wastes, heavy metals, etc. As in the case of Thailand above, the facilities for dealing with these are presently inadequate.

The factors which contribute most to atmospheric pollution include carbon monoxide, hydrocarbons, oxides of nitrogen and sulfur, polyaromatic hydrocarbons, chloroflurocarbons, and aerosols. Certainly the increase in the size of Taipei has led to an increase in air pollution, though there are apparently problems of measurement. It is possible to control the damage from e.g. car emissions, though the technology has been spreading more slowly in Taiwan than in Japan or the US because of cost. Cars are also responsible for much of the noise pollution which Williams sees as a major problem in Taipei.

The other obvious results of population growth and industrial development in Taiwan are pressure on basic resources: agricultural land, water, and wildlife and biodiversity. As a result of urbanization, agricultural land has begun to shrink as the population has continued to grow. There are also problems of soil erosion on slopes, and the pollution of agricultural land near to cities and industrial developments. Also a problem is the use of chemical fertilizers. The net result of all these processes is that there is less land and it is becoming less and less productive.

It may seem strange that a country with heavy rainfall should suffer from water shortages, but increasingly Taiwan, just like Thailand, is experiencing a shortfall in supply. Barely 25% of the runoff from rain is used because of seasonal variations in rainfall, and limited reservoir capacity. The reservoir system is also silting up due to soil erosion. This means that the cities are dependent on ground water resources, with the usual problems of subsidence and salt water coming in.

As for diminishing biodiversity, Taiwan, like Thailand, is trying to cope with the problem through extension of the national parks system, though in the case of Taiwan there is now very little wildlife left to protect, despite a growing conservation movement in recent years.

If the pattern of Taiwan's industrialization has much in common with that of Japan and Korea, so does its environmental movement. As with Korea, for much of the postwar period Taiwan had an authoritarian government which stifled protest, though this began to change after the death of Chiang Kai-shek in 1975 and the accession of his son, Chiang Ching-kuo. From that period onwards the political system has been liberalized, and it has also begun to respond to what it sees as public concern about the environment.

NGOs have become active, have published journals, and have served as pressure groups for change. Even so, as Williams notes, it is difficult to say how large the movement is as a whole, and many people still seem more concerned with their own prosperity than the island's environment. In a couple of well publicized cases public opposition to local development schemes actually resulted in their cancellation or other concessions by the government.

One was the proposal to build a Du Pont titanium dioxide plant in Lukang, an old port city on the west coast which aroused opposition despite assurances that it was safe. The local residents were afraid that it would harm the local aquaculture.

A second case was in Houchin, with the proposal to extend an existing petrochemical plant there. The scheme was eventually put into operation, but only after the company had promised to install the latest pollution control technology.

A third example was another petrochemical plant in Ilan, a fishing town in the north east. The scheme involved one of Taiwan's leading industrialists, and he was forced by the strength of the local protests to debate the issue with the local governor on television.

There have been many other examples both of protests and in some cases direct action by local citizens groups, either directed at new developments or demanding compensation for old ones. The government response was to create a Bureau of Environmental Protection in the early 1970s and legislation to prevent water pollution was introduced in 1974. Generally the Bureau achieved little, though exceptions are the cleaning up of the Love River, a heavily polluted area near Kaohsiung Harbor, which was pushed by Chiang Ching-kuo himself, and which provided a model for similar schemes elsewhere.

The government also began to require environmental impact assessments before development projects could be implemented. In the 1980s, the establishment of the Environmental Protection Agency with a bigger budget showed that the government was putting a higher priority on environmental protection in order to tackle the environmental mess. One of the worst examples is the Tamsui river which flows from the coast through Taipei and which has for years been a dumping ground for human, animal and industrial waste. This will require the upgrading of the Taipei sewerage system and the provision of more landfill sites for garbage, as well as a ban on dumping by private business and individuals. The sheer volume of garbage generated by Taiwan's growth also means that the government is having to start incinerating it, as is now usual in Japan, but which is likely to increase worries about dioxin.

Another cause for environmental concern in Taiwan is the country's nuclear energy program, based on six reactors constructed since the late 1970s which supply about half the country's electricity. The two main questions have been what to do with the waste, and how safe is the system, after a fire in one of the reactors in 1985 (the year before Chernobyl). This led to strong opposition to the construction of yet another plant. Even though the scheme has only been postponed rather than actually canceled, it does show that the environmentalists have been successful in taking on some of the biggest operators in the Taiwanese economy and causing them to defend or rethink their plans.

Williams' summary of the Taiwan experience is succinct. First, as it demonstrates in relation to environmental problems, prevention is much cheaper than cure. However, in most rapidly developing countries, the first priority is rapid growth rather than environmental protection, partly as a result of political pressure for a higher standard of living. Environmental initiatives are, therefore, most likely to be taken as a result of activism by a minority of private citizens, prodding the government into action. But they are only likely to be successful if the government cooperates, as it is probably only the government which can convince the rest of the population that protecting the environment is important not only for quality of life, but for economic growth. Economic growth, in turn, will have to include safeguards for the environment, with controls on industries and commodities which are highly polluting. As one example, the motorcycle may have to be phased out in cities if emissions cannot be controlled. In Taiwan's case cleaning up the mistakes made over the last thirty years is daunting, with an EPA staff of 318 trying to monitor 70,000 factories, over 8 million vehicles, and a population of 20 million. Education and use of the media is, therefore, a vital component of its long-term strategy (Williams, 1994: 252-53).

China

China presents even greater problems, which have been best analyzed by the Canadian scholar, Vaclav Smil in two books written at an interval of ten years, *The Bad Earth* (Smil, 1983), and *China's Environmental Crisis* (1993). It is his second book that I mainly draw on here.

His starting point is the crisis of population pressure, especially in the east of the country. Smil notes that even using the most optimistic statistics, the population of China is likely to grow by 125 million during the 1990s, a number equivalent to the population of Japan. By 2025 there will be another 200 million or more, close to the current population of Indonesia. The country of course has a long history of violent population fluctuations due to famine and war, but these are now less likely than in previous periods.

The one-child policy has not stopped population growth, and the Chinese population will probably not stabilize for some years. When it does so it will stabilize at a higher level than it is at now, before starting to fall. This will increase the already intense pressure on the agricultural land in the east of the country. Much of this is already affected by soil erosion, desertification, and pollution on an even larger scale than the other countries discussed above. In recent years these processes have been made worse by the increasing use of chemical rather than organic fertilizers. China has generally been successful in feeding its population during the period of economic growth that started in the late 1970s, but the present pattern of agriculture cannot really be sustained.

As in Taiwan, but on a much larger scale, there is also a water shortage. The densely populated north of the country, in the Beijing-Tianjin area and the three Manchurian provinces to the north where a lot of the country's industry and 40% of the population are located receive only 25% of the rain, and 10% of the available river water. Beijing, like other major Asian cities, has a major water shortage. The pumping of ground water has led to severe subsidence: this is the major constraint on the further expansion of the city, and could eventually lead to the relocation of the capital. Much of the water goes into the irrigation system in agriculture, and then simply runs away and is wasted. Clearly it could be much better managed, but this costs money and with the breakdown of the commune system it is now probably more difficult than it would have been in the 1970s.

Further economic development will require more water, but it is not clear at the moment where it will come from.

A further result of the population pressure has been the disappearance of the forests over most of the country, despite a program of reforestation since the 1949 revolution. The provinces where the forest is still concentrated (Sichuan, Yunnan, Hainan) have lost between a third and two thirds of their trees in a period of 40 years, and the area of remaining forests in the country as a whole decreased by a third in the five years from 1982-87, as the economic growth speeded up. The present rate of wood consumption is not sustainable, and there is a good chance of all the forests disappearing in the near future.

One reason why the trees are disappearing of course is that the local people need them for cooking and heating. Many parts of rural China have a chronic and serious energy shortage. The main energy source is coal, but the transport and distribution systems are both poor. Also very little of the coal is washed before it is burned, so it causes a lot of air pollution and acid rain, which kills more of the forest. The current concern about carbon dioxide will mean that China will be pressured in future to cut its use of fossil fuels, but it is not clear at the moment where the additional energy it will need for further development will come from.

As China continues to grow economically, the increased demand for housing, water, food, energy, and raw materials can only make the environmental problems worse before it gets better. Smil's grim conclusion is that it is probably impossible to reverse these trends within the next decade. Many Chinese will face an increasing struggle to survive, and the environmental problems will have a negative impact on the country's development as a whole.

But one hopeful sign which he does see is that these issues are being more openly discussed in China now than they were before, and have resulted in both increased public awareness of the problems and legislation to protect the environment. And this takes us to the final part of the discussion: the political processes through which environmental improvements might or might not come about at some point in the future.

Environmental crisis and political response

In this brief survey of the economic growth, environmental problems and planning processes of some of the major East and Southeast Asian economies, there are substantial similarities in the experience of the different countries in the region. In all of them, the postwar period was marked first by a drive for economic growth, followed by a realization of the environmental problems which this growth was causing. In most cases, development plans and guidelines, under the pressures of environmental and social movements have begun to reflect environmental concerns, and environmental protection agencies have been set up. The relationship between economic growth, environmental movements and the state is shown in Figure 1.

How much impact these changes have had, however, varies considerably, and in this final section I will attempt to summarize some of the differences.

In the case of China the drive for rapid economic growth which began in 1958 known as the Great Leap Forward was a major disaster, and the period of sustained economic growth only really began twenty years later, with Deng Xiaoping's economic reforms from 1978 onwards. Although many of the negative impacts of this growth on the environment are already only too clear, the emphasis of government policy is still on rapid economic growth. For a large part of the population, the environmental problems have been offset by rising living standards and life expectations, and large parts of the country are still waiting for the early benefits of high speed growth to appear, in the form of better roads, communications and electrification. Even in the more wealthy cities, the main political problems which the government faces is that of protests from the workers laid off in the unprofitable state enterprises, and of government workers in general affected by inflation, which is seen as directly under government control. Some of the more obvious environmentally desirable policies, such as raising the price of energy to both private consumers and industry are simply impossible to carry out in the present political climate. China also relies to a large extent on its own capital resources, and so in the case of major projects like the Three Gorges Dam, they have pushed ahead despite protests from domestic and foreign environmentalists alike.

It will be clear from Smil's analysis discussed above, that he thinks that the Chinese face great difficulties in moving beyond this environmental impasse, and that it places great constraints on the country's future economic and political development. A more optimistic scenario is that presented by William Overholt in his book *"China: The Next Economic Superpower* (1993)". He argues that already China has experienced a degree of economic growth unparalleled for so many people at any time in history. He also believes that generally this growth is likely to continue, makingChina

Stage	Government priority	Type of regime	Environmental movements	Examples
First Stage/ High-speed growth	High-speed growth of basic industry and incomes	Authoritarian/ military or one-party state	Dormant/Repressed	China
Second Stage/Society moving to advanced industrialism	Economic growth based on high technology and services	Paternalistic government, closely linked to business interests	Nascent, mainly involving local and minority cultural or ethnic groups concerned with effects on local resources	Thailand, Indonesia, Malaysi
Third Stage/ Advanced industrial society	Mitigation of environmental damage	Increasing openness and democratization	Growing strength and middle class support, concerned with environmental impact of industry on quality of life	Korea, Taiwan
Fourth Stage/ Post-industrial society	Restoration of damaged environment, development of leisure infrastructure	Plural, including parties representing environmental interests and capital	Established environmental movements, concerned with impact of new service and leisure industries on biodiversity etc.	Japan

Figure 1. Relationship between economic growth, environmental movements and the state.

the largest economy in the world early next century, and that it will bring with it eventual political change. As in Korea and Taiwan, the changes will come about because of the creation of a large, wealthy middle class which will increasingly demand political representation, and the result will be an increasingly pluralistic political system. At that point presumably the interests of consumers will be increasingly taken into account, and these will include environmental and quality of life issues, in addition to basic provision of food, housing and work.

It would be good if Overholt rather than Smil turns out to be right. Probably the actual situation will be somewhere between the two. Despite Overholt's observation that the interior areas of China in the early 1990s were actually growing faster than the coastal belt, it may still happen that China in the next century will be rather polarized into a rich coastal zone and a poor underdeveloped interior. With the growing power of the provinces, some of the richer ones such as Fujian and Guangdong (with access to capital from Taiwan and Hong Kong) or the metropolitan areas of Shanghai and Beijing-Tianjin (the main centers of administration and financial services) may well develop rather quickly economies looking rather like those of China's richer neighbors Korea, Taiwan and Japan. Growth in the land-locked provinces of the interior will probably be generally rather slower, and environmental issues there will take a lot longer to resolve.

Even in these provinces, it may also be that what will develop will be a political system in which power lies in the hands of the political elite and the capitalist class, and that, as in Indonesia, Thailand, and Malaysia, it will be the wishes of these groups rather than an environmental constituency which will prevail. The problems of monitoring the environment in Thailand and Indonesia will remain, and Malaysia will increasingly put more and more stress on the kinds of resort development which have long been a feature of Japan, but which are also damaging to the environment. Eyal Ben-Ari (1998) has described well they ways in which the leisure industry, especially golf, is spilling over the border from Singapore into Indonesia and Malaysia.

Korea and Taiwan seem to have moved further in the direction of taking environmental issues more seriously, perhaps because the level of environmental damage is greater due to their more highly developed economies. Unlike Thailand, Malaysia, and Indonesia described by Boyle(1998), the Taiwanese environmental movement seems more able to take on and extract concessions from local capitalists and politicians, and Taiwan also seems to have realized that prevention is much cheaper than cure when it comes to dealing with environmental issues. In these countries, with a large, wealthy middle class and basic problems of housing, employment and infrastructure already solved, environmental issues have come to the top of the agenda for a large section of the population, and politics may increasingly reflect this in the next few years. Clearly, the more successfully an environmental component is included in the education of the younger generation in all these countries, the higher the level of environmental awareness in the adult population, and the quicker these political processes will come about.

There is also an international dimension to solving the problem. A recent comparison of environmental impact assessment (EIA) by Boyle in Malaysia, Thailand, and Indonesia distinguishes between projects which are initiated by local capital and those initiated by capital from outside. He shows that in all three countries EIA faced substantial resistance and limited support from public- and private-sector leaders and decision-makers in the case of purely domestic projects. On the other hand in cases where foreign funds were involved, and where international attention was focused on the environmental impact of the project, EIA was given much more emphasis (Boyle, 1998: 104-6).

A final glance at Japan, however, provides something of a cautionary note to this on which to end. McKean's book was written in the late 1970s and early 1980s, and the political message was that people were standing up for their rights, and that environmental issues were coming to the fore in local politics, just as they have in Korea and Taiwan according to Kim and Williams. Twenty years on, little has actually changed. A glance at two recent analyses of Japanese politics published in English makes clear the kinds of processes going on and the difficulties of changing the system.

Of particular interest are the various analyses by foreign scholars of "Japan Inc." and especially the construction industry and its links with government (McCormack, 1996; Woodall, 1996). The government provides a steady supply of work at high prices to the construction industry, which in turn supports the ruling party and provides comfortable jobs for retiring bureaucrats. In the 1950s and 1960s the main construction projects were industrial. In the 1970s and 1980s many of them were connected with leisure and services: airports, golf courses, and so on. Many of them also have long-term consequences for the environment, even if this is less obviously harmful than the industrial projects of the earlier period. In the case of the various projects around Tokyo and Osaka Bays, these more recent projects have been very large and very expensive, even if the economic rationale for some of them (e.g. an airport at Kobe, following the opening of the Kansai airport between Osaka and Wakayama) is not entirely clear. In many cases the protests of environmental groups against these projects continue to be ignored. Most analyses suggest that in fact the limits to these projects are not environmental, but financial: as the deficit of the state rises, and as bubble economy style projects become increasingly uneconomic, protests will grow and politicians and bureaucrats will be forced to become more realistic, even if it does mean alienating the large construction firms.

However, the construction industry and the process of "rebuilding the Japanese archipelago" as former prime minister Tanaka Kakuei described it, cannot be written off that easily. The fact is that many of these construction projects are actually financed by private capital, and many of them are financially viable. Other recent books show well just how adaptable Japanese capital has been in defining consumption and peoples' tastes, and selling them what they believe that they want. Most successful of all has been the Tsutsumi family who has taken the lead in the development of retailing, transport, sports (both baseball and winter sports), hotels and leisure resorts (Havens, 1994), but there are many others who have followed their lead, and the results have been seen not only in Japan but in the resorts, golf courses and department stores of the rest of the region, from Beijing to Jakarta and beyond.

This highlights two basic contradictions at the heart of the world capitalist system.

First, planners and bureaucrats can plan, but things only happen when people with capital appear ready to invest it, so their perceptions of whether they will make a profit, and how much, have to be taken seriously by politicians and officials. So far all the evidence from Asia is that it is their views that matter, whether or not they are environmentally friendly.

Second, politicians may have their own views and ideals in relation to the environment, but they also need to generate the cash to get into office and stay there, and the current system in Japan provides them with this. If change is to come about it will have to be through reforms which make it in the interests of politicians and their careers to support environmental issues, rather than simply to pay lip service to them in their speeches and ignore them in reality.

Ultimately, therefore, in a post-industrial economy such as Japan, the future of the environment may depend on how far it is in the interests of capital and the political leaders to be environmentally friendly, and at the moment the political discourse is still all about improving life by improving services, infrastructure, and transport, which means more concrete and more damage to the environment. Japan is not alone in facing this problem, but until the advanced industrial societies of the world do, the chances of major improvements to the environment will probably remain relatively bleak.

References

Ben-Ari, E. 1998. Golf, organization, and "body projects": Japanese business executives in Singapore. In:*The Culture of Japan as Seen through its Leisure*. Linhart, S., Früstück, S. Eds, SUNY Press, Albany, New York, pp 139-61.

Boyle, J. 1998. Cultural influences on implementing environmental impact assessment: insights from Thailand, Indonesia, and Malaysia. In:*Environmental Impact Assessment Review*, 18: 95-116.

Havens, T.R.H. 1994. *Architects of Affluence: The Tsutsumi Family and the Seibu-Saison Enterprises in Twentieth-Century Japan*. Harvard University Council on East Asian Studies, Cambridge Massachusetts.

McCormack, G. 1996. *The Emptiness of Japanese Affluence*. M.E. Sharpe, Armonk, New York.

McKean, M.A. 1982. *Environmental Protest and Citizen Politics in Japan*. University of California Press, Berkeley, California.

Overholt, W. 1993. *China: the next economic superpower*. Weidenfeld and Nicolson, London.

Smil, V. 1983. *The Bad Earth*. Zed Press, London.

Smil, V. 1993. *China's Environmental Crisis: An Inquiry into the Limits of National Development*. M.E. Sharpe, Armonk, New York.

Smith, R. 1978. *Kurusu: The Price of Progress in a Japanese Village, 1951-75*. Stanford University Press, Stanford, California.

Williams, J.F. 1994. Paying the price of economic development in Taiwan: environmental degradation, in *The Other Taiwan: 1945 to the Present*, Rubenstein, M.A. Ed., M.E. Sharpe, Armonk, New York, pp 237-56.

Woodall, B. 1996. *Japan under Construction: Corruption, Politics, and Public Works*, University of California Press, Berkeley.

Chapter 2

Lake Biwa and the Asian Environmental Agenda: Issues and Prospects

Masahisa Nakamura, Lake Biwa Research Institute, 1-10 Uchidehama, Otsu, Shiga, 520-0806, Japan.

The Evolving History of Lake Biwa

Lake Biwa is located in the upper part of the Yodo River along which lie such major cities as Kyoto and Osaka.[1] The lake's watershed closely coincides with the area of Shiga Prefecture where the local residents are known to have developed throughout history a very special attachment to the lake. The man-nature interaction involving the lake and its surroundings has also been quite unique and extensive, making the Omi people (the people in the lake region) develop a unique sense of association with the lake.

Among the distinct phases of Lake Biwa management, the earliest and the longest of the phases was that of management of floods and droughts. The paddy farmers in the catchment basin had periodically suffered from these natural disasters.[2] This phase was followed several decades later by that of management of water resources. When Japan began to exhibit phenomenal economic growth in the 1960s, the downstream water demands began to grow rapidly, leading the downstream prefectural and municipal governments to express a strong interest in the withdrawal of additional amounts of Lake Biwa water. After lengthy political jockeying among officials representing Shiga Prefecture, the downstream local governments (municipal and prefectural), and the central government (the Ministry of Construction), the framework of the Lake Biwa Comprehensive Development Project (LBCDP) emerged and was finally agreed upon in 1972. The basic idea of the LBCDP was to allow the discharge of a maximum of 40 m^3/sec. of additional water in times of droughts which amounts to the water supply needs of several million people. The project was comprised of water resource development works, flood control and related water management works, and the compensatory public works for Shiga Prefecture. It was originally a ten-year project, but not all of the component projects were completed by 1982, and the project duration was extended for another ten years. It was further extended in 1992 for an additional five years and was finally completed in March, 1997. The total budget of this 25-year national project, the LBCDP, is reported have surpassed 1.8 trillion yen, or about 13 billion dollars.

Upon completion of all of the LBCDP component projects, the additional amount of lake water can be released to meet the downstream needs at times of severe droughts. The improved river and the coastal levee systems will help alleviate the possibility of flood damage. In addition, Shiga Prefecture, which was relatively underdeveloped before the commencement of the project, has become much better off in terms of its economic prosperity realized in part through the LBCDP. At the same time, it has been transformed from a mainly agricultural to a mainly industrial prefecture.[3] There have been gradual shifts in industrial structure from primary to secondary, and secondary to tertiary.[4] The future potential of the region has also been greatly improved, with extensive infrastructure development undertaken thanks to the financial arrangements made through the LBCDP.

The management of Lake Biwa today, however, is about to enter a new phase, a phase which will be devoted to environmental conservation, and particularly to the restoration of the whole of the lake ecosystem. A brief review of the history of lake water quality management is in order.

The Evolving History of Lake Biwa Water Quality Management

The first ominous sign of lake water quality deterioration came, all of a sudden, in the form of a large scale red tide outbreak along the eastern coastline of the Northern Basin of Lake Biwa in early 1977.[5] Though it was the policy of the Prefecture not to allow the sitting of severely polluting industries to begin with, the discharge of polluting effluents from industries was still a major concern in the 1970s. By the mid-1980s, however, the stringent regulatory measures by the government had drastically reduced the industrial contribution to lake pollution. Also prompted by the red tide incident, Shiga Prefecture enacted the Eutrophication Control Ordinance of 1980 to ban the use and sale of synthetic detergents containing phosphorus. This was made possible thanks primarily to the initiatives, efforts, and support coming from the citizens, most significantly in the form of housewives' movements.

Though the control of land-based sources of pollution had progressed to some extent before then, it was only after this particular period that the Prefecture began to succeed in mobilizing resources for undertaking extensive water quality management control measures. A wide array of control measures has been introduced over the decades for improving Lake Biwa's water quality, and some of them are notable for their significant contribution to improving water quality, including stringent regulation of pollution from industrial effluents. While the LBCDP originally did not include lake water quality as being a primary ingredient of the plan, it turned out to be closely associated with water quality issues both with respect to the eventual inclusion in the plan of component projects for upgrading water quality, and with respect to alleviation of the negative impacts on lake water quality of the development projects and activities directly or indirectly related to LBCDP.[6]

Household waste water has been, and still is, a major polluter, the treatment of which requires various forms of sewage treatment systems. Construction of these systems for municipalities was slow at the beginning; but after the first extension of the LBCDP in 1982 the pace picked up significantly. The regional public sewerage system is planned to cover more or less the entire flat-land part of the watershed.[7] There is also a significant contribution from non-point sources of pollution like those from rainwater, forests and agricultural fields including paddy fields, and urban open spaces.[8] The introduction of a comprehensive non-point source control system will be expensive and will require legal and institutional measures yet to be elaborated.

The evolving processes of development and conservation taking place over the past few decades around Lake Biwa together show the changing trends in the quality of lake water and of the lake ecosystem. The typical water quality indices show that, despite all the environmental programs implemented, the improvement in lake water quality has not been so impressive.[9] In addition, there are other more subtle indications in that there is a dramatic change in the dominant species of nuisance-causing plankton in the lake. To summarize, it may be stated that the trend towards deterioration accelerated up till the mid-1980s, and, though the rate of deterioration has been slowed down since then, the trend itself has not been reversed quite yet.[10] The management of Lake Biwa is about to enter a new phase which will require greater commitment to improve the water quality.

The Shaping of the Lake Biwa Environmental Agenda

Although the Lake Biwa issues described above pertain only to the management of water resources and lake water quality, other aspects of the environmental agenda have also been a vital component of the evolving process of lake management. For example, broad categories of issues such as waste management including discharge and control of atmospheric pollutants, municipal solid wastes, and hazardous and toxic chemicals have been intertwined with the management of liquid wastes containing eutrophication-causing nutrients. The planning and management of such land resources as urban, agricultural, and rural, as well as forest lands have also been of vital importance in the shaping of the state of the lake environment. Further, there have been important human dimensions, such as the prefectural demography as affected by the regional demography and the improvement of per capita prefectural income status, as well as human health and settlement status. With regard to the evolving issues relating to the Lake Biwa environment and the experience of lake management in this respect, some observations of importance may be made as follows:

Notable characteristics of lake environment deterioration

There are two important facts to recognize in the deterioration of lake environments, i.e.;

1) The deterioration of lake water quality progresses slowly and steadily taking a long period of time, yet it is extremely difficult to predict its sudden acceleration and the dramatic alteration of the ecological system which can take place in a relatively short period of time.
2) It is extremely difficult to rehabilitate the lake ecosystem to the original water quality and ecological balance once its degradation reaches a threshold level.

In the case of Lake Biwa, the lake is likely to have been undergoing steady change over a long period of time. It was only a little more than two decades ago that the deterioration of water quality was recognized as a serious problem. With all the managerial, institutional, and technological countermeasures taken in the past years, the quality improvement has been slow and the prospect of recovery appears uncertain.

Dealing with conflicting lake management issues

The communities interacting with a lake are often forced to deal with two basically conflicting objectives, e.g., watershed development vs. protection of fisheries, or withdrawal of quality water out of the lake and the discharge of its return flow back into it. How the structure of lake management problems is shaped depends on how the water use and discharge processes evolve. In addition, the societal sense of the value of water, be it water quantity or water quality, changes over time. In the case of Lake Biwa, the consensus on the importance of the lake among the basin residents and the downstream users has gradually emerged after many years of political battles and of partial resolution of conflicts of interest.

The Evolving Process of Lake Management

There are always multiple means to adopt in countering lake water quality deterioration, be they hard technological solution approaches, like sewage and industrial waste water treatment, or soft institutional and managerial ones, including legislative actions, public involvement and environmental education. It is not that these pieces fit together nicely at the outset but they gradually reinforce each other. Some were designed to function as a part of administrative system while others arose more spontaneously, and the system evolved by itself. Flexibility and spontaneity are essential ingredients in the pursuit of multiple countermeasures.

As stated above, the water quality restoration of highly exploited lakes requires mobilization of a great deal of technical input and financial resources. It is rare also to find a single countermeasure that fulfills all of the restoration requirements. Multiple measures usually evolve over decades and in stages. The first set of measures sought in general is technical and structural. The limitation of the approach is quickly recognized principally for financial reasons. The second set of measures is institutional. Regulatory measures are effective to the extent of the enforcement capability of the managing body of the lake. The conflicting interests in lake resource utilization may also defy the expectation of successful implementation. The third set of measures is more social. Often it takes the form of resolution of conflicts and collaboration by many segments of the community in the region.

The restoration activities are conducted in many different forms and on various degrees of scale. Public involvement is one of the key ingredients at this stage. Many new issues arise in the evolving process of restoration measures, as the responses to countermeasures are varied and unexpected. A reasonable balance of many factors contributing to a solution will eventually emerge in the process, and the persistent and continuous pursuit of multiple measures will eventually pay off.

Parallel Reference to the Asian Environmental Agenda

In view of the complexity of environmental management having been experienced by Lake Biwa and the downstream communities, it would be interesting to make some reference to the similarly complex situation in Asia involving development and the environment.

Features of the Asian environmental agenda

The countries of Asia, particularly Southeast Asia, are said to have succeeded over the past decades in transforming their national economic structures from ones heavily dependent on primary industries to ones based on secondary and even tertiary industries. The region has been boasting national economies with very high rates of growth for some time. Accompanying this, however, was the rapid transformation of these basically rural and agricultural regions into urban and industrial societies.

Development is inherently self-contradictory. As the national economies expand, the gap in living standards between the rich and the poor becomes wider. Conflicts of interest related to the use of air, water, and land resources arise between different sectors of national economies, between different industrial sectors, between different generations, and between different geographical regions. The conflicts of interest in the use of water resources are particularly severe between the upstream and downstream users, and between the urban sector and the rural agricultural sector, not to mention the competition over use among different industrial sectors. In addition, there are social and cultural impacts. Most notably, the social ties which would have necessarily been quite strong in the rural community setting may be gradually eroded, becoming less important in urban and industrial societies.

The resulting issues regarding the environment, land, water, and air, are varied, many and intertwined. After the global summit of 1992 in Rio de Janeiro, these issues have been laid out in Agenda 21 together with new environmental issues such as global warming, the loss of bio-diversity, the depletion of tropical forests, and the growing hazardous waste generation caused, in part, by sitting polluting industries from overseas. The examination of Agenda 21 in the implementation record of the Rio Conference has led to the issuance of the Rio+5 statement. It reconfirms the need for greater effort to achieve sustainability in development, to pursue an integrated approach in the management of environmental resources, to attain higher efficiency in the mobilization of financial resources, and to enhance the level of citizen participation.

The parallelism with the Lake Biwa experience

As in the case of Lake Biwa, the Asian environmental agenda in general has been shaped by allowing development to take place for the enhancement of regional and national welfare. In the initial stage of development, it was possible to allow development activities to take place without causing environmental degradation beyond the level of local concern. As long as the magnitude of resource extraction activities was maintained within a certain threshold level, it was possible to keep the degree of environmental resource degradation at a minimum. It was, thus, possible to maintain a coherent social system which suitably catered for this development — environmental dynamics of moderate intensity.

As a nation aims to transform itself from a rural and agricultural state to an industrial one, the scale of economic development has to be expanded, and the level of resource extraction activities will be pursued beyond the limits of self-renewal. The degradation of environmental resources then proceeds in much more profound ways than may be apparent. By then, pressure for development is perpetuated both from within and from outside; from within, because of the need to catch up with the industrialized nations in terms of economic achievements, and from outside because of the need to accommodate to the globalizing economy.

As the economic development pressure mounts, exploitation and degradation of environmental resources accelerate basically as a result of the development activities and also to accommodate development needs. Such a situation may be tolerated for some time, but it soon becomes apparent that the degradation of environmental resources leads in many cases to the loss of economic opportunities. Thus environmental restoration becomes a major concern both for the environmental and the development sectors. The restoration of the environment will not be easy because of the financial resource constraints. In general, development activities are bound to precede the environmental enhancement activities, and the amount of financial resources required for restoring the original environmental capacity would be enormous.

Conflicting views exist also about sharing financial resources, e.g., between investor and host nations concerning trade, and among different sectors within a nation on sharing the economic gains generated. The societal values are also changing dramatically, particularly with regard to the rising expectation for achieving higher standards of living. As consumerism becomes widespread among the general public, environmental awareness may not keep pace, as reflected in, for example, the ever-growing amount of solid wastes, not to mention emissions into the air and water of noxious pollution.

The process of adopting multiple countermeasures has progressed, albeit with limited success in bringing about tangible results. Technical and structural means have been and are being pursued, mostly through overseas technical collaboration. The requirement of enormous financial resources has been hampering their progress in most of the Asian nations, as expected. The institutional measures are beginning to take effect for resolving certain issues, such as the attainment of ambient environmental quality in air and water. To further strengthen the institutional measures, however, human resources development programs have to be continually updated and improved, which takes time and resources as well. The social approach has yet to mature to the extent of producing tangible results in most nations, though there are growing number of environmental NGOs playing an extremely important role which cannot be fulfilled by any other institutional systems.

If the Lake Biwa experience provides any general clue to the productive pursuit of environmental issues, it is, among others, that environmental considerations have to be integrated within development activities, that prevention is better than cure, that resource use has to be properly directed, and that multiple countermeasures have to be sought and brought into the development scheme with the involvement of all sectors of society.

Conclusion

The Lake Biwa experience provides parallels with many environmental agendas of other countries. They all face growing pressure to balance development with conservation, urban and industrial amenities with rural values and traditions, and human service requirements with ecosystem requirements. This is typically the case with rapid transformation in many of the faster developing regions of Asia. This unlikely parallelism has, in fact, been proven to be quite illuminating in the transfer of the experience gained from

Lake Biwa to many, including those who have come, for example, to share their concerns in many different forums for international collaboration.

Notes

1 Lake Biwa, the greatest lake in Japan and one of the greatest lakes in Asia, is 675 km^2 in surface area and 27.5 billion m^3 in volume. As a single water body supplying water to some 14 million people, it can also be regarded as one of the most important water resources in the world today. The lake is believed to have come into existence about five million years ago, and together with Lake Baikal and Lake Tanganyika, it is known to be one of the oldest lakes in the world. Just as is the case with other ancient lakes, Lake Biwa provides a habitat for many indigenous fauna and flora. The lake, therefore, is one of the most important water bodies in terms of freshwater bio-diversity and is an important subject of scientific research.
2 The completion of weir construction in 1905 across the only outflowing river, the Seta River, has since moderated the magnitude of their impacts.
3 The traditional textile industries of marginal scale, for example, were joined by large textile mills around the turn of the century. They were subsequently replaced by more profitable manufacturing operations ranging from machinery to electronics.
4 Today their shares are, respectively, 2.4%, 57%, and 34.6%.
5 The sighting of the phyto-plankton, *Uroglena americana*, was quite a shock to the Shiga residents as the northern basin of Lake Biwa had been, up till then, believed to be nearly pristine. The red tide has since been sighted almost every year.
6 The negative impacts of LBCDP became a major issue, for example, in a 13 year-long litigation brought against the prefectural and national governments by a group of citizens claiming their rights to preserve the lake environment without having it subjected to extensive alteration as specified in the plan. Though the litigation ended in 1989 with the plaintiffs losing the case, it drew a great deal of attention and later affected, to a significant degree, the thinking about the development of the post-LBCDP measures.
7 It currently serves some 30% of the entire population (which corresponds to a little over 40% of the planned service population), with other smaller scale treatment systems. The service coverage of the public sewerage system is currently

expanding by a few percentage points a year, and is, thus, expected to reach around 70% coverage by around 2010.
8 The financial resources required to construct and manage such facilities will be enormous. Proper management of irrigation water and reduction of wasteful use of fertilizers and pesticides are the keys to successful control of non-point source runoffs.
9 Also, the water transparency and the amount of oxygen at the bottom of the northern basin of the lake show that Lake Biwa water quality had slowly been deteriorating even before the red tide incident of 1977.
10 The emerging features of control measures characterizing the new phase include 1) realignment of protected watersheds and land uses, 2) development of ecotone areas including restoration of the once reclaimed attached lakes, and 3) the integrated management of priority watersheds.

Selected Bibliography

Nakamura, M. 1995. Lake Biwa: have sustainable development objectives been met? *Lakes and Reservoirs: Research and Management.* 1: 3-29

Nakamura, M. 1996. The emerging Lake Biwa issues: from water resources development to restoration of ecological environment. *International Shiga Forum on Technology for Water Management in the 21st Century, Shiga Prefecture and UNEP-ITTC, 25-27 November 1996.* Otsu, Japan, 43-54.

Nakamura, M. 1997. Preserving the health of the world's lakes. *Environment.* 39(5): 16-20, 34-40.

Chapter 3

Beyond Environmental Legislation: Environmental Education in Malaysia

Sham Sani, Department of Geography, Penerbit University Kebangsaam Malaysia, 43600 UKM Bangi, Selangor, D.E. Malaysia.

Introduction

This paper argues that no conservation program, however well it may be designed, can be completely successful without public support which can only come from a well-informed citizenry. To create such an environmentally committed society, a carefully thought out awareness and education program, both formal and informal, needs to be in place. It is essential that formal environmental education be complemented by an awareness program which is best undertaken as a joint effort between government and non-governmental organizations (NGOs), the media, and the private sector.

The present paper is specifically concerned with Malaysia although the argument applies equally well to other societies. It provides an overview of the consequences of rapid development on the environment and the institutional efforts made to resolve and manage these consequences, a discussion of the need for an effective environmental education and awareness program as an alternative, an assessment of the existing program, and future challenges.

Economic Achievements

Like many countries of the developing world, Malaysia placed a high priority on economic development, and for some years after independence in 1957 this was even at the expense of almost everything else, including the environment. Through a series of successive five-year plans, the economy expanded considerably, especially during the 1970s and the 1980s.[1] The Gross Domestic Product (GDP) grew at a rate of 6% per annum in the 1960s, almost 8% per annum in the 1970s, and 6% per annum in the 1980s. Rubber and tin no longer dominated the economy; and by the end of 1990, the manufacturing sector accounted for 27% of the GDP compared with about 14% in 1970. Following the recession in the mid-1980s, development policies and strategies were reviewed. Economic policies placed stronger emphasis on growth, structural adjustments, and the liberalization of the economy. Privatization was given strong emphasis. Liberal investment policies were introduced in order to accelerate the growth of private investment and private entrepreneurship. Following a decline of about 1% in 1985, the economy recovered strongly after 1987. A growth rate of 8.5% was achieved in 1989 and 9% in 1990 (Ministry of Finance, 1990: 17). As the economy entered the 1990s, the prospects of becoming a Newly Industrializing Country (NIC) appeared to be promising. In his 1996 Budget address to Parliament, the Minister of Finance, Dato' Seri Anwar Ibrahim, noted that the Malaysian economy continued to be strong (Anwar Ibrahim, 1996). The GDP growth increased at an average rate of close to 9% per annum for eight consecutive years and was projected to grow at an average rate of 8% per annum with low inflation in 1997.

The last decade saw a tremendous transformation of the economy with the manufacturing sector contributing an increasingly significant percentage to the GDP. As a comparison, in 1960 the manufacturing sector contributed a mere 8.5% to the GDP. This figure increased to 20.5% in 1980, 26.8% in 1990, 31.5% in 1994 and 33.1% in 1995 — well beyond the year's target of 32.7% (quoted from Anuwar Ali, 1995: 11). The structure of Malaysia's export commodity market also underwent a tremendous change as the contribution from manufacturing increased from 59% in 1990 to 68.6% in 1992 and to 80% of total exports in 1995 with a corresponding decline in agriculture from 20% in 1990 to 11% in 1995. Similarly, employment in the manufacturing sector also increased at an average rate of 8.9% per annum. Its share of total employment increased from 15.8% in 1984 to 25.5% in 1995. The share of manufactured exports relative to merchandise exports likewise increased from 11.1% in 1970, to 20.6% in 1980, to 58.8% in 1990, and to 79.6% in 1995.

Environmental Consequences of Development

While development achievements since independence have been creditable and impressive, such a rapid pace of development has not been without detrimental effects on the natural environment. This is not surprising as the basis for Malaysia's growth and development has been its relatively rich natural resource base, both renewable and nonrenewable. Basically, there are two major issues that need redress: the depletion of resources and the deterioration of the environment.

Development and resource depletion

An important resource which has been greatly affected in the pursuit of economic development is the rain forest. At the end of the Sixth Plan (1995), the total land area under forest was only 59% (*The Seventh Malaysia Plan*, 1996-2000). While the figure compares favorably with those of many other countries, there has been an increased concern that the rate of forest depletion is excessive. In Peninsular Malaysia, for example, forest cover steadily declined from about 69% of the total land area in 1966 to 55% in 1978, and 47% in 1990 (Lee, 1973; Othman Abdul Manan, 1991). Some authors quoted an even lower figure (see Potter, 1993: 105). Large-scale agro-conversion of forested land, the construction of hydro-dams, mining, commercial logging, and shifting cultivation are the major causes of forest depletion. Tho (1991) estimated that the total forest areas that have been flooded as a result of dams being built in the country could be well over 100,000 hectares, not including the Pergau Dam and the more than 90,000 hectares flooded in the construction of the Bakun Dam in Sarawak.

Another resource which is currently being affected by development is biological diversity for which Malaysia is particularly well-known. Indeed, Malaysia has been identified as one of the 12 "megadiversity" countries of the world. It contains some 185,000 species of fauna and about 12,500 species of flowering plants. Malaysia is also well endowed in plant genetic resources. Following rapid development, there has been a general reduction and loss of biological diversity largely through the loss of genetic resources, flooding as a result of dam construction, deterioration in the quality and quantity of water supply, decline in the food supply, and loss of productive soils and potentially useful biological resources. Such reduction is particularly noticeable in coastal and marine resources. The rapidity with which development is taking place in the coastal zone together with poor planning and design of coastal development projects has given rise to many problems related to sustaining a balanced development. Economic pressures have led to the destruction of the mangrove forests to make way for aquaculture, agriculture, and tourist resort development, while exploitation of coastal resources in excess of a sustainable level has caused serious depletion in fish catches. In addition, rapid industrial and infrastructural development in the hinterland have also contributed to increased organic and inorganic pollution of rivers and coastal waters. The destruction of mangrove and other wetland forests, which serve as breeding grounds for fish and prawns, has subsequently caused a decline in fish resources.

Like the other resources, those of energy and minerals are also being exploited to sustain the current rate of development. Despite the National Mineral Policy in 1992 and the promulgation of the Mineral Development Act 1994, there is always concern about overexploitation of these resources and the environmental degradation they cause.

Development and environmental degradation

Apart from resource depletion, development also affects the environment. Air and water are the two major aspects of the environment that have been seriously affected in recent years. Toxic and hazardous wastes together with municipal solid wastes are also fast becoming a problem that needs to be addressed.

The three major sources of air pollution, especially in urban areas, are mobile (motor vehicles), stationary (power stations, industrial fuel burning processes, and domestic fuel burning), and the burning of municipal and industrial wastes. In 1995, they contributed almost 100% of the pollutant emissions discharged into the atmosphere. In addition, there was also an increase in the incidence of acid rain and haze. The latter was particularly serious in 1991, 1992, 1994, and more recently in September-October 1997. The 1994 Environmental Quality Report (DOE, 1995, 12-13) noted that while air quality in the country for 1994 was generally good, urban centers like Kuala Lumpur, Ipoh, Johor Bahru, and Kuching showed an increase in the level of total suspended particulates in the air. It was also observed that, on average, the hourly concentrations of respirable particulates (< 10 um diameter) during the peak hazy periods were 4-6 times higher than normal.

In terms of river quality, the Department of Environment (DOE) reported that, overall, based on the Water Quality Index (WQI),[2] there has been a slight decline during the Sixth Plan (1991-95) period. Of the 119 rivers monitored, 52 were classified as "clean", 53 as "slightly polluted", and 14 as "highly polluted". The number of highly polluted rivers increased from six to fourteen in 1994 (DOE, 1995). Sewage, agricultural, and industrial wastes respectively contributed 65%, 27% and 8% of water pollution in terms of BOD. Highland developments and activities in the inland areas contributed an increase in suspended solids that caused an increased incidence of flooding as well as coastal and marine pollution.

With regard to coastal and marine pollution, and apart from the contribution of suspended solids from development activities upstream, oil spills formed a major problem. For the 1991-95 period, some 146 major oil spills were reported: 51 in the Straits of Melaka and 95 in the South China Sea.

Solid waste disposal also poses a serious problem especially in urban areas. Following an increase in the number of people who live in urban areas and allowing for 0.34-0.85 kg/capita/day of waste generation in urban areas, the problem of solid waste disposal and the availability of disposal sites is expected to increase. Unsafe landfills and illegal dumping of waste will continue to remain a problem. Apart from household and solid wastes, Malaysia also handles a wide range of hazardous and toxic chemicals. The *Seventh Malaysia Plan* noted an estimated hazardous waste generation of 337,000 tons for 1992. Major industrial sources include metal finishing, electrical and electronics, food processing, textiles, chemicals, and iron and steel manufacturing.

Management Responses

The depletion of resources and the deterioration of the environment, especially over the last quarter of a century, have caused a great deal of concern for the public and the government. Control measures and other forms of responses are essential in order to manage and conserve the environment and the natural resources.

Perhaps an early form of management response to impending environmental problems and depletion of resources was through legislation. In this regard, the process began as early as 1894 when the Straits Settlement Ordinance No. 3, 1894, which protected several species of wild birds, came into being. This was followed by several other environment-related pieces of legislation, including the Waters Enactment in 1920, the Mining Enactment in 1929, the Forest Enactment in 1934, the Drainage Works Ordinance in 1954, the Road Traffic Ordinance in 1958, the Land Conservation Act in 1960, the Fisheries Act in 1963, the Factories and Machinery Act in 1967, and the protection of Wildlife Act in 1972.

Currently, some 40 environment-related pieces of legislation in Malaysia are being enforced (Table 1). It needs to be emphasized, however, that many of these were not originally designed specifically to address environmental problems, but rather to promote sound practices in specific sectors in line with government policies. In addition, many of the pieces of legislation enacted prior to the 1970s

Table 1. Environment-related Legislation in Malaysia

1	Waters Enactment 1920
2	Mining Enactment 1929
3	Mining Rules 1934
4	Forest Enactment 1935
5	Natural Resources Ordinance 1949
6	Poisons Ordinance 1952
7	Merchant Shipping Ordinance 1952
8	Sales of Food and Drugs Ordinance l952
9	Dangerous Drugs Ordinance 1952
10	Federation Port Rules 1953
11	Irrigation Areas Ordinance 1954
12	Drainage Works Ordinance 1954
13	Medicine (Sales and Advertisements) Ordinance 1956
14	Explosives Ordinance 1958
15	The Road Traffic Ordinance 1958
16	Land Conservation Act 1960
17	National Land Code 1965
18	Housing Development Act (Licensing and Control) 1965
19	Radioactive Substances Act]968
20	Civil Aviation Act 1969
21	Malaria Eradication Act 1971
22	Continental Shelf Act 1966 (Revised) 1972
23	Petroleum Mining Act 1972
24	Environmental Quality Act 1974
25	Geological Survey Act 1974
26	Street, Drainage and Building Act 1974
27	Aboriginal Peoples Act 1954 (Revised) 1974
28	Factories and Machinery Act 1967 (Revised) 1974
29	Pesticides Act 1974
30	Destruction of Disease-bearing Insects Act 1975
31	The Protection of Wildlife Act 1972 (Revised) 1976
32	Antiquities Act 1976
33	Local Government Act 1976
34	Town and Country Planning Act 1976
35	National Parks Act 1980
36	Malaysian Highway Authority Act 1980
37	Pig Rearing Enactment 1980
38	Atomic Energy Licensing Act 1984
39	Exclusive Economic Zone Act 1984
40	National Forestry Act 1984
41	Fisheries Act 1985

were largely sectoral, focusing on specific activity areas. Extensive as they were, sector-based legislation did not encourage an integrated approach to environmental policy implementation, neither were they able to cope with the increasingly more complex environmental problems that we face today. Consequently, the Environmental Quality Act (EQA) was conceived.

The EQA has been described as the most comprehensive piece of legislation concerning environmental management in Malaysia. It was passed by Parliament in 1974. The spirit embodied in this Act was officially endorsed in the *Third Malaysia Plan* (1976-80) and continued to be the thrust of the *Fifth Malaysia Plan* (1986-90), the *Sixth Malaysia Plan* (1991-95), and the *Seventh Malaysia Plan* (1996-2000). The fundamental need for sound environmental management in planning and implementation of development programs, as contained in the *Third, Fifth, Sixth,* and *Seventh Malaysia Plans*, provides the guiding principles for the National Environmental Policy objectives.

Essentially, the 1974 EQA forms the basic instrument for achieving the National Environmental Policy objectives. The Act provides for an advisory Environmental Quality Council (EQC) whose function is generally to advise the Minister-in-Charge of Environment on matters pertaining to the Act and those referred to it by the Minister. The Act also provides for the appointment of a Director-General (DG) of Environment whose duties and functions include the issuing of licenses for waste discharge and emissions, the formulation of standards, the coordination of pollution and environmental research, and the dissemination of information and educational materials to the public.

To assist the DG, a Division (now Department) of Environment (DOE) was established in 1975. In the administration of EQA, a three-pronged strategy was developed by the DOE as follows:

1) To control pollution and take remedial actions;
2) To integrate an environmental dimension in project planning and implementation; and
3) To provide environmental inputs into resource and regional development planning.

Under the first strategy, several regulations were promulgated for enforcement (Table 2). To date, at least 20 regulations have been promulgated under the EQA in 1974. Among the latest to be introduced were those relating to toxic and hazardous wastes.

One area where legislation has been successful in reducing the pollution load is in the production of crude palm oil. Figures reported by DOE (1990) show that since the enforcement of the Environmental Quality (Prescribed premises) (Crude Palm Oil) Regulations in 1977, the total revenue collected in licensing the

Table 2. Environmental pollution control regulations gazetted under the EQA 1974.

	Regulation/Order	Effective Enforcement Date
1.	Environmental Quality (Prescribed Premises) (Crude Palm Oil) Regulations 1977 Amendment (1982), P.U. (A) 342	November 4, 1977
2.	Environmental Quality (Licensing) Regulations 1977, P.U.(A) 198.	October 1, 1977
3.	Motor Vehicle (Control of Smoke and Gas Emissions) Rules 1977 (made under the Road Traffic Ordinance, 1958), P.U.(A) 414	December 22, 1977
4.	Environmental Quality (Prescribed Premises) (Crude Palm Oil) Order 1977, P.U. (A) 199.	July 1, 1978
5.	Environmental Quality (Prescribed Premises) (Raw Natural Rubber) (Amendment) Order 1978, P.U. (A) 337	April 1, 1978
6.	Environmental Quality (Prescribed Premises) (Raw Natural Rubber) Regulations 1978 (Amendment 1980), P.U. (A) 338.	December 1, 1978
7.	Environmental Quality (Clean Air) Regulations 1978, P.U. (A) 280.	October 1, 1978
8.	Environmental Quality (Compounding of Offenses) Regulations 1978, P.U.(A) 281.	October 1, 1978
9.	Environmental Quality (Sewage and Industrial Effluents) Regulations 1979, P.U. (A) 12.	January 1, 1979
10.	Environmental Quality (Control of Lead Concentration in Motor Gasoline) Regulations 1985, P.U. (A) 296.	July 11, 1985
11.	Environmental Quality (Motor Vehicle Noise) Regulations 1987, P.U.(A) 244.	July 16, 1987
12.	Environmental Quality (Prescribed Activities) Environmental Impact Assessment Order 1987, P.U. (A) 362.	April 1, 1988
13.	Environmental Quality (Scheduled Wastes) Regulations 1989, P.U.(A)139.	May 1, 1989

14.	Environmental Quality (Prescribed Premises) (Scheduled Wastes Treatment and Disposal Facilities) Order 1989 P.U.(A) 140.	May 1, 1989
15.	Environmental Quality (Prescribed Premises) (Scheduled Wastes Treatment and Disposal Facilities) Regulations 1989, P.U.(A) 141.	May 1, 1989
16.	Environmental Quality (Delegation of Powers on Marine Pollution Control) Order 1993, P.U. (A) 276.	September 23, 1993
17.	Environmental Quality (Prohibition on the use of Chlorofluorocarbons and other gases as Propellants and Blowing Agent) Order 193, P.U. (A) 434.	October 25, 1993
18.	Environmental Quality (Delegation of Powers on Marine Pollution Control) (Amendment) Order 1994, P.U. (A) 536.	December 29, 1994
19.	Environmental Quality (Delegation of Powers on Marine Pollution Control) Order 1994, P.U. (A) 537.	December 29, 1994
20.	Environmental Quality (Prohibition on the use of Controlled Substances in Soap, Synthetic Detergent and other Agents) Order 1995, P.U.(A)115.	April 15, 1995

industry has declined by about 88% by 1989. The revenue collected decreased from M$2,768,000 in 1978 to M$714,000 in 1980, and M$338,000 in 1989, reflecting the extent of the reduction in pollution loads discharged by this industry. A similar success, albeit less dramatic, was observed in the raw natural rubber industry. The total revenue collected in licensing the raw natural rubber industry since 1979 shows a reduction of 44% in fees collected by 1989 (DOE, 1990).

While the main objective of the first strategy is basically to ensure that the existing industries and other pollution sources are subject to direct controls or "add-on" technologies, such remedial measures alone, without the support of some form of preventive controls, are inadequate. In view of this, an environmental impact assessment (EIA) procedure was introduced as an integral part of the overall project planning. The Environmental Quality (Prescribed Activities) (Environmental Impact Assessment) Order of 1987 was enacted and enforced on April 1, 1988. Unlike pollution control legislation, EIA is essentially a preventive measure and an aid to the environmental planning of new projects or to the expansion of existing ones. It was designed basically to identify and predict the magnitude of the environmental impact of proposed projects so that adverse environmental effects might be avoided. Early identification of likely impacts also enable environmental experts to study in depth a limited number of impact areas which are most likely to be of major significance.

In the integrated project planning approach, environmental considerations are incorporated into project planning and implementation via a mandatory requirement under Section 34A of the Environmental Quality (Amendment) Act of 1985. This section requires anyone who intends to undertake a prescribed activity to first conduct a study to assess the likely environmental impact from the activity and the mitigating measures that need to be taken in order to overcome it. Under the Environmental Quality (Prescribed Activities) (EIA) Order, 1987, the proposed project cannot be implemented until the EIA report is approved by the DG of Environment. The Environmental Quality (Prescribed Activities) (EIA) Order, 1987, specifies some 19 categories of activities requiring EIA reports prior to project approval for implementation.

A third strategy adopted by the DOE is to incorporate environmental inputs into resource and regional development planning such as regional, master, structure, and local plans. The DOE has been proactive in this respect and plays a catalytic role through its participation in the planning of projects by various government agencies. It has also supported this through the supply of environmental data and information besides the various guidelines which are distributed to the public (Table 3). In order to enhance the three broad strategies described above, a number of support programs have also been developed. These include:

1) environmental monitoring and assessment;
2) environmental education, information, training, and public awareness;
3) environmental research and development;
4) inter-agency and federal-state cooperation;
5) program coordination through the state environment committees; and
6) bilateral, regional, and international legal and institutional arrangements.

The extent to which these programs are being affected depends greatly on available funding allocated to the DOE. The

Table 3. Some Environmental Guidelines for Planners and Project Developers

1	Guidelines for the Sitting and Zoning of Industries
2	Guidelines for the Disposal of Solid Wastes on Land
3	Guidelines for the Prevention of Soil Erosion and Salutation
4	Guidelines for Toxic and Hazardous Wastes Management
5	Guidelines for Environmentally Sensitive Areas in Malaysia
6	A Handbook of Environmental Impact Assessment Guidelines
7	EIA Guidelines for Industrial Estate Development
8	EIA Guidelines for Golf Course Development
9	EIA Guidelines for Coastal Resort Development Projects
10	ELA Guidelines for Municipal Solid Waste and Sewage Treatment and Disposal Projects.
11	EIA Guidelines for Development of Resort and Hotel Facilities in Hill Stations.
12	EIA Guidelines for Mines and Quarries
13	EIA Guidelines for Dams and/or Reservoirs Projects
14	EIA Guidelines for Fishing Harbors and/or Land Based Aquaculture Projects.
15	EIA Guidelines for Drainage and/or Irrigation Projects
16	EIA Guidelines for Thermal Power Generation and/or Transmission Projects.
17	EIA Guidelines for Ground water and/or Surface Water SUDPIY Projects.

monitoring program, for example, would do well with some extra allocation to upgrade instrumentation and related facilities, although certain aspects of monitoring and assessment (e.g. air quality) are already being privatized.

Figure 1 gives a summary of the environmental management strategy currently adopted by the DOE. It highlights the following three main elements of the strategy:
1) pollution control, EIA and land use planning;
2) the role of the states through the State Action Committees on Environment; and
3) the support programs.

In the first 13 years of its establishment, the DOE concentrated almost exclusively on pollution control with some limited effort on preventive measures. Following the enactment of the Environmental Quality (Prescribed Activities) (EIA) Order, 1987, which became effective on April 1, 1988, and the efforts to ensure sustainable development, the preventive approach in environmental management was emphasized.

Not all aspects of the environment are adequately covered by the 1974 EQA or the Environmental Quality (Amendment) Act, 1985. Issues such as forestry, water resources, mining, wildlife, and fisheries are beyond the jurisdiction of the EQA and the DOE. Under the Malaysian Constitution, land, for example, is a state matter. Each state is empowered to enact laws on forestry and to formulate forest policy independently. In order to facilitate a coordinated approach, the National Forestry Council (NFC) was established in 1971 by the National Land Council (NLC). The National Forestry Policy, which was accepted by the NFC and later endorsed by the NLC in 1978, formed the basis for forestry management in the country. Other environmental issues which are not covered by the EQA are subject to separate legislation administered by specific government agencies in cooperation with the DOE.

Environmental Awareness and Attitudes

A great deal has been achieved over the last 20 years since the DOE was established. The Environmental Quality (Prescribed Activities) (EIA) Order, 1987, which enabled the EIA procedures to be legally implemented, was an important step forward. The tremendous improvement in palm oil and rubber waste control management, the reduction of lead in petrol from 0.4 gm/l prior to 1990 to 0.15 gm/l on January 1, 1990, and the increased public awareness of environmental issues are only some of the achievements in Malaysia's efforts to conserve and enhance environmental quality. In fact, in terms of lead reduction in petrol, all the major petroleum

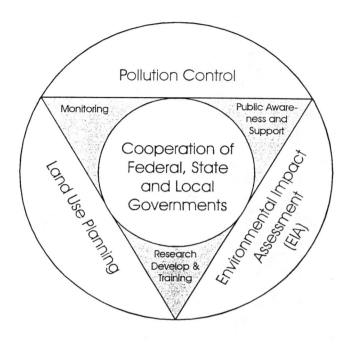

Figure 1. Environmental Management Strategies

companies operating in Malaysia have now gone a step further by making unleaded gasoline available to the public. The move from a largely "curative" management stance during the first 13 years of the EQA to a more "preventive" approach has been very encouraging. The successful efforts to get the states to establish the State Action Committee on Environment and the EPU to agree to the concept of a "National Conservation Strategy" have also been achievements. All of these successes will undoubtedly encourage greater faith in future environmental conservation in Malaysia. In the meantime, however, there is still a great deal more to be accomplished.

One persistent issue which appears to have caused concern within environmental management circles is public awareness and attitudes. A study of public environmental awareness conducted by the DOE some 10 years ago (DOE, 1986) suggests considerable public apathy toward environmental issues. As a public issue, it was ranked second to last. Over the last five years, the situation appears to have changed considerably. The increase in the number of public complaints about environment-related problems and the increase in media coverage of environmental issues suggest an increased awareness. However, based on the nature and extent of environmental degradation today, one wonders if the level of awareness is sufficient to ensure a healthy future for the environment. Furthermore, awareness alone is not likely to be effective unless it is also accompanied by a commitment to environmental improvement. The bottom line is that no conservation program, however well it may be designed, can be completely successful without public support. The latter can only come from well-informed citizens who are aware and fully committed. This includes all sections of the community from administrators, politicians, and the private sector right down to ordinary people in the street and school children.

Obviously, one important policy consideration with respect to environmental strategies in the 1990s is the need to increase the level of public awareness and commitment to the environmental cause. While legislation and institutions that administer policies and programs of environmental management are important, public support is equally essential in order to ensure their success. At both the federal and state levels, efforts to educate the public and disseminate environmental information must be intensified. However, because environmental education is basically aimed at engendering community actions, a varied approach involving government institutions, the media, the private sector, and NGOs are needed to reach the different target groups.

The role of the NGOs in environmental education has long been acknowledged. Their activities are aimed at effecting changes and shaping attitudes. Both directly and indirectly, they are involved in environmental education. Of great significance is the role of the NGOs in providing a mechanism for feedback to the government and its regulatory agencies on the negative side-effects of program implementation. In many respects, NGOs are the public watchdogs for the proper use of natural resources, conservation, professional practices and other activities of the government and the private sectors which adversely impinge on the environment. The government, on the other hand, should be willing to listen to alternative views without prejudice. In some cases, this can be difficult as environment and development are very often closely linked and environmental NGOs may easily be dismissed as being anti-development or trouble makers.

Education for the Environment

Recognizing that learning is a life long process, public education for the environment needs not just the schools and universities but also learning in the workplace and in the community. It is here that central and local governments, schools, universities, businesses, NGOs, and community organizations can play a role. A number of key concepts should be part of the environmental education program and should be pitched according to the audience. Some of these key concepts include an understanding of:
1) Ecosystems and ecological principles;
2) The links between environment, economy and society;
3) Sustainable use of resources;
4) The system of environmental management (e.g. Malaysia, Indonesia, Thailand);
5) Different attitudes and values in relation to the environment;
6) An ethic of stewardship or guardianship; and
7) Individual and collective responsibility for the environment.

Priority areas for environmental education include:
1) Schools;
2) Teacher education;
3) Business and professional education;
4) Cleaner production techniques for business;
5) Action-oriented programs in the community; and
6) Access to information about environmental education programs, courses and resources.

In Malaysia, many of these imperatives are already being incorporated and put in place in the formal education sector such as schools and universities. In the first six years of primary education, for example, students are introduced to the basic elements of the natural ecosystems in subjects like local studies and general science. Later, in secondary schools, more varied environment-related subjects like geography, general science, biology, chemistry, and physics are offered. These are complimented by some environment-related activities organized under the "co-curriculum" component, which is compulsory in all schools, both at the primary and secondary levels.

At the university level, as in many countries, courses and programs in environmental science, management, and conservation are offered up to the post-graduate level. Research is also being undertaken in the different aspects of conservation. Indeed, some universities have even established specialized research institutes in environment and development. The efforts at the university, however, are confined to a limited number of students as environment-related courses and programs are only some of the many they need to choose from.

As for the public, both the government (especially through agencies like the DOE, Department of Forestry, Department of Fisheries, the Department of Wildlife and the Department of Agriculture) and the NGOs have regular, organized programs in the form of seminars, teach-ins, environmental campaigns, and projects. The universities also get involved in the NGO activities as well as those of the government agencies. Such involvement has enabled the universities to provide the public with the latest research findings and development in the environmental sector.

One important focus of the NGOs' activities is the school children's involvement in environmental awareness programs. Exhibitions, talks, organized trips, and study camps are some of the activities planned for primary and secondary school children. Sometimes these are organized in collaboration with the DOE, the Department of Wildlife, the Department of Forestry, or even the corporate sector.

Table 4 lists some of the major environment-related NGOs in Malaysia, their dates of establishment, and size of membership. It will be observed that apart from the Malayan Nature Society (MNS) which was established in 1940, the rest of the NGOs are relatively small. Obviously, more support is needed from the public in order to create strong NGOs. However, despite these constraints the NGOs with support from the government and the private sector have successfully undertaken several conservation projects that are open to the public.

The MNS, for example, is particularly active and has several education field centers such as the Kuala Selangor Nature Park, the Endau-Rompin Reserve, and the education center at the Forest Research Institute in Kepong near Kuala Lumpur. Many of these centers are run in collaboration with a government agency, the state government, or the private sector.

The Kuala Selangor Nature Park in Kuala Selangor, for example, is a "joint venture" between the MNS and the Selangor State Government. It is made up of a variety of habitats including secondary forest (201 ha), mangrove forest (95 ha), the estuary of the Selangor River, and mud flats and brackish water lake systems,

Table 4. Some major environment-related NGOs in Malaysia

Organization	Date established	Membership (approx.)
MNS (Malayan Nature Society)	1940	3000
FOMCA (Federation of Malaysian Consumer Assn.)	1973	
EPSM (Environment Protection Society Malaysia)	1974	140
WWF (Malaysia) (World Wide Fund for Nature)	1972	
CETDAM (Centre for Environment, Technology and Development)	1985	50
ENSEARCH (Environmental Management and Research Association of Malaysia)	1984	600
CAP (Consumer Association Penang)	1969	
SAM (Sahabat Alam Malaysia)	1977	500+
CETEC (Centre for Environment and Technology)	1992	
APPEN (Asia Pacific People's Environment Network)	1983	
PEM (Persatuan Ekologi Malaysia)	1992	

with more than 150 species of birds, insects, reptiles, and mammals, including silvered leaf monkeys and leopard cats. A nature trail system has been established to enhance the study of flora and fauna of the area. Research activities within the area are currently being conducted in collaboration with local universities and research institutes. The Nature Park is open to the public and regular programs are organized for school children.

Concluding Remarks

This paper draws attention to the fact that an environmental education and awareness program is an essential component of any environmental management strategy. In Malaysia, a number of such programs are already in place although these can be expanded considerably. Both the government and the NGOs are already involved in many of the current environmental education programs. The NGOs, however, need more support from the public so that their watchdog function can become more effective and relevant. As many of the NGOs are operating on a limited budget, the contribution from the private sector can go a long way to enhance their environment-related programs.

Notes
1 Since the formation of Malaysia in 1963, there have been seven successive Five-Year plans as follows:
   First Five-Year Plan (1966-70)
   Second Five-Year Plan (1971-75)
   Third Five-Year Plan (1976-80)
   Fourth Five-Year Plan(1981-85)
   Fifth Five-Year Plan (1986-90)
   Sixth Five-Year Plan (1991-95)
   Seventh Five-Year Plan (1996-2000)
2 The Water Quality Index (WQI) appraises water quality based on five parameters, i.e. biochemical oxygen demand, (BOD), chemical oxygen demand (COD), ammoniacal nitrogen, suspended solids, and hydrogen levels.

References
Anuwar Ali. 1995. *Globalisasi Pembangunan Industri dan Peranan Pemerintah Di Malaysia. Syarahan Perdana Jawatan Profesor UKM, 18 Nov. 1995.* [Globalization of industrial development and the role of Government in Malaysia. Inaugural Lecture, UKM, 18 Nov. 1995]. University Kebangsaan Malaysia (UKM) Press, Bangi, Malaysia.
Anwar Ibrahim (Finance Minister). 1996. The 1997 Budget Speech, at the Dewan Rakyat. Malaysian Parliament, Kuala Lumpur, Malaysia.
DOE. 1986. Public Environmental Awareness. Unpublished report Prepared for the Department of the Environment by Frank Small Associates, Kuala Lumpur, Malaysia.
DOE. 1990. *Environmental Quality Report 1989.* Department of the Environment, Kuala Lumpur, Malaysia.
DOE. 1991. *Environmental Quality Report 1990.* Department of the Environment, Kuala Lumpur, Malaysia.

DOE. 1995. *Environmental Quality Report 1994*. Department of the Environment, Kuala Lumpur, Malaysia.

Lee, P.C. 1973. Multiuse management of West Malaysia's forest resources, In: *Biological Resources and National Development,* Soepadmo, E. and Singh, K.C. Ed. Malayan Nature Society (MNS), Kuala Lumpur, Malaysia, pp. 93-101.

Ministry of Finance. 1990. *Economic Report 1989-90*, Ministry of Finance, Malaysia, Kuala Lumpur, Malaysia.

Othman Abdul Manan. 1991. Forest resources in Malaysia. In: *Proceedings of the National Seminar on Environment and Development, Kuala Lumpur, Malaysia, July 9 - 11, 1990.* Department of the Environment, Kuala Lumpur, pp. 173-225.

Potter, L. 1993. The onslaught on the forests in Southeast Asia, In: *Southeast Asian's Environmental Future. The Search for Sustainability*, Brookfield, H. and Broom, Y. Ed., UN University Press and Oxford University Press, Kuala Lumpur, Malaysia, pp. 103-123.

Tho, Y.P. 1991. Conservation of biodiversity: international and national perspectives. In:*Proceedings of the National Seminar on Environment and Development, Kuala Lumpur, July 9 - 11, 1990.* Department of the Environment, Kuala Lumpur, Malaysia, pp. 266-313.

Chapter 4

Mountains, Rivers, and Coastal Zones in Asia: Environmental Management and Community Integration into the 21st Century

Azra Meadows and Peter S. Meadows, Biosedimentology Unit, Division of Environmental and Evolutionary Biology, Institute of Biomedical and Life Sciences, University of Glasgow, Glasgow G12 8QQ, Scotland, UK.

Introduction

Mountains, rivers, and coastal zones in Asia represent one of the most important interrelated environmental systems on a global scale, particularly in terms of predictions of climate change and global warming into the 21st Century (IUCN 1990; Houghton et al. 1996; Watson et al., 1996; Batterbury et al., 1997). It is interesting, therefore, that until very recently few scientists and even fewer politicians realized their central role in controlling the use of land and water resources by human populations (Dobby, 1962; Groombridge, 1992). The implications for the biodiversity and resource potential of a country cannot be over emphasized (Table 1). Huge areas of natural habitat have already been lost on a global scale. In Asia, these range from 41% in Malaysia to 80% in Vietnam (Table 2). Recent international political initiatives such as the 1992 Rio Earth Summit on Biodiversity, the 1997 Earth Summit II in New York, the Kyoto Climate Summit in December 1997, together with international scientific conferences such as the Indus River meeting in 1994 (Meadows and Meadows, 1998a), show that at last the political and scientific communities of developed and developing countries are becoming aware of the future dangers and threats to their countries and are beginning to act. Nowhere is this more important than in the Asiatic region, where research and education in these areas is vital.

The Asiatic region is central to the future development and prosperity of *Homo sapiens* as a species. It has potentially huge resources and some of the most important natural habitats and world heritage sites, but the region also has major problems of human population increase and pollution. The region contains some of the most highly populated countries, including India, Bangladesh,

Table 1. Natural and human impacts on mountains, rivers, and coastal zones.

Type of Environment	Impact	Process	Output	Environmental Management
Mountains	Natural impact	Global warming, melting of glaciers, weathering of rocks	Water discharge, soil erosion	Prevention of buildup of greenhouse gases
	Human impact	Deforestation, erosion	Soil erosion, silting	Reforestation, managed land use
Rivers	Natural impact	Flooding	Soil particulates, salts, water discharge	Flood control defenses, dams, barrages, dikes
	Human impact	Waterlogging, desertification, salinization, industrial and domestic pollution	Abandoned agricultural land. Pollutant discharge, water quality	Construction of outfall drains. Pollution control
Coastal Zones	Natural impact	Global warming, sea level rise, hurricanes, cyclones	Flooding of low lying coastlines. Loss of human life, habitat.	Coastal defenses. Weather predictive models, satellites.
	Human impact	Industrial and domestic pollution. Over-exploitation of natural resources. Habitat degradation.	Decline in fisheries. Eutrophication. Depletion of natural resources. Loss of nursery grounds for fish.	Pollution monitoring. Managed resource exploitation. Conservation, protection.

Table 2. Habitat loss (area and percent) in some Asian and African countries.

Country	Total Area (km2)	Area lost (km2)	Percent Area lost
Indonesia	1,446,430	708,751	49
South Africa	1,236,500	704,805	57
Ethiopia	1,101,000	770,700	70
Burma	774,820	550,122	71
Madagascar	595,200	446,400	75
Cameroon	469,400	276,946	59
Malaysia	356,250	146,063	41
Vietnam	332,100	265,680	80
Ivory Coast *	318,000	251,220	79
Philippines	308,200	243,478	79

*Cote d'Ivoire
(Source: IUCN and UNEP, 1986a, b; WRI et al., 1992).

Pakistan, China, and Indonesia. It is also a major area of tectonic activity, which has resulted in the development of huge mountain ranges, episodic and destructive seismic activity, earthquakes and hurricanes. At the same time it contains great biogeographical diversity both in terms of ecosystems and species of animals and plants.

The Himalayan-Karakoram-Tibet mountain complex with its unique high mountain environments, and rivers such as the Brahmaputra, Ganges, Indus, Mekong, and Yangtze that originate in these mountain ranges are unique. These mountains and rivers are all in Asia. This, together with the presence of some of the most biologically diverse and globally important coastal zone mangrove swamps and coral reefs, means that Asia is of central significance to the whole global ecosystem.

The history and prehistory of *Homo sapiens* and even the origins of our civilizations have depended on access to water and to cultivable land, and these are associated with mountains, rivers, and the coastline (Andrew, 1986; Wheeler, 1968, Possehl, 1979; 1993). In Asia, in addition to the often isolated communities in the mountains, much trade has involved crossing massive mountain ranges (Bashir and Israr-ud-Din 1996). The historic silk route across the Himalayas is a classic example. Mountains, rivers, and the coastal zone are, therefore, intimately bound up with the development of human populations, and they will continue to determine how and where we live for the foreseeable future.

There is an additional reason for considering mountains, rivers, and the coastal zone as one interacting system. The various parts are entirely interdependent in a geomorphological and environmental sense (Figure 1).

Mountain erosion is a natural process by which flash floods and mountain streams feed water and sediment into the main parts of a river catchment area. Periodic river floods distribute water and fine sediment onto the flat country of the terrestrial basin of a river, thus producing good agricultural land. It is not often appreciated that the soil cultivated by present day man has been transported there from mountainous areas. This soil has been transported by the continuous action of melting glaciers and rivers, and by the floods that frequently occur during the monsoon season in Asia. Mountain streams and rivers themselves are sources of irrigation for surrounding land, and have been used as such for thousands of years. The "falaj" system, first developed by the Persians, is an old irrigation system of canals used in the Sultanate of Oman. The irrigation of the Indus Basin, first developed during the Moghul period, was much expanded during the British occupation of the area in the 19th Century and first half of the 20th Century. Finally, much of the land in the coastal zone is formed by historical or present day river estuaries and deltas, often operating on a massive scale. Towards the coastline itself, large areas of sand dunes and mangrove swamps are dependent on river flow, and the effects offshore of river sediment transport can be detected in large submarine deltaic fans that stretch for many hundreds of kilometers from the coastline (Milliman et al. 1984; Meadows and Meadows, 1998b).

Mountains, rivers and the coastal zone have therefore to be considered as a series of contiguous interacting ecosystems, in which water and sediment are transported from high mountain environments, onto river basins to form rich agricultural land, and

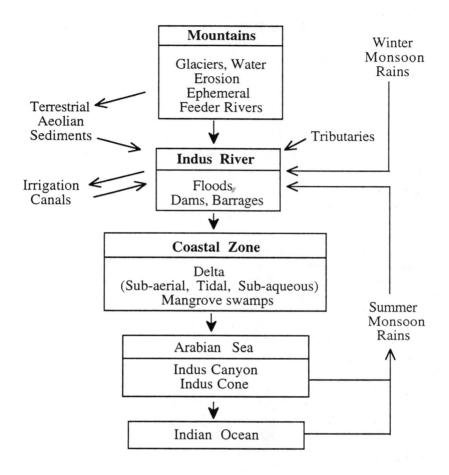

Figure 1. Mountains, the Indus River and the coastal zone as an interacting system: an Asian Example.

then to the coastal zone and into the oceans. The countries of Asia contain many such mountain, river, and coastal zone systems.

Mountains

Mountain environments are extremely variable partly because of wide ranges of altitude, temperature, and rainfall, partly because of different vegetation, soil, and sediment types, and partly because of different levels of human population (Gerrard, 1990; Ahmed and Sheikh 1994). Farming usually consists of small holdings which are terraced on steep slopes. In some areas farmers have holdings at different levels that allow them to use different plant species and animals (Hawtin and Mateo, 1990). In the Afghanistan area, for example, farmers move to higher land in the summer season and return to lower ground in autumn. Villages also often have communal pastures and croplands. In Nepal, the ripening of wheat and barley occurs approximately five days later with each 100 m increase in height. Above the frost line land is more usually used only for grazing. Other problems are associated with the wide variation in temperature between day and night when differences of 20° to 30°C are encountered. There are also very large seasonal differences. In some cases summer temperatures of 30°C can be followed by temperatures of -30°C in winter.

There are inherent difficulties associated with the steepness of much mountain agricultural land. All of this means that it is not easy to apply directly agricultural practices that have been developed for farming on low lying plains to high altitude mountain environments (Byerlee and Husain 1992; Riley et al., 1990). Having said this, the diversity of plant species exhibited by mountain farming areas is very high. Hawtin and Mateo (1990) report that in one Nepalese village at least 150 species or varieties of plant are being cultivated, and small holdings in the High Andes, Peru, can contain 12 different species of potatoes. However, much of this diversity and its genetic basis is in danger of being lost. Some of the causes are urban demand, increasing population pressures, and the introduction of new varieties with high yields.

There are also many naturally occurring species of animals and plants that appear to be endangered in mountain regions, although scientific information is often scanty. In the Hindu Kush and Karakoram ranges to the west of the Himalayas, a number of mammal and bird species are recorded as endangered or threatened (NWFP/IUCN, 1996; Roberts, 1991; 1992; 1997). Endangered species include the Chinkara gazelle, flat horned and straight horned markhor, snow leopard and brown bear (mammals) and the Cheer pheasant, Western Tragopan pheasant, Houbara bustard, peregrin falcon, Saker falcon and Siberian crane (birds). The latter four species of birds are migratory. There is less information about plant species (Baquar, 1995; Nasir et al., 1995). In this context it is particularly pleasing that a UNDP supported program involving

IUCN in the Chitral District and Northern Areas of Pakistan is adopting a community based approach to conservation of species such as these (NWFP/IUCN, 1996, p 160). The program is based on work started by WWF and the Northern Areas Administration of Pakistan in the Bar Valley area. Communities in three valleys will receive training in wildlife management. They will also be requested to identify and then to manage selected conservation areas. In exchange for this work the communities will receive a percentage of revenues accruing from any sustainable future programs in the selected areas. There are also other initiatives. It is hoped that private game reserves will be established, in which 75% of the revenue will be retained by the landowners. All of these initiatives are excellent, and should serve as models for community based cooperation in wildlife management in other mountain regions in Asia.

Soil types in mountain environments are extremely variable although they are often stony and only provide a thin cover over rock. There is also a need for soil maps for most areas using modern techniques such as geographical information systems (GIS) and satellite imagery (Burrough, 1986; Soeters et al., 1991; Rengers, et al., 1992; Price and Heywood, 1994). Perhaps the most important feature of most mountain environments is the potential for soil erosion by glaciers, rain and seasonal torrents (Embleton, 1972; Gerrard, 1990). Many rural communities in these areas are in danger of over-exploiting their environment, partly because of increasing contact with the outside world. There is a balance between the community itself and its use of natural forests and water on the one hand, and its modification of suitable slopes or (Figure 2). Natural forest and alpine pastures below the snow line flat ground for terraced agriculture or cultivated fields on the other provide fuel, wood for building, and grazing areas for livestock, and the livestock in turn provide the community with food, wool and hidea. The balance is a delicate one that can easily be upset either by changing practices within the rural community itself or by the intrusion of outside business wishing to capitalize on underdeveloped natural resources. A typical example is the timber Mafia that is said to operate in northern Pakistan.

Anyone who has worked at high altitude knows that plant cover which inhibits soil erosion is minimal above about 3000 m. We have observed many steep slopes in the Chitral, Gilgit and Hunza regions of the Hindu Kush and Karakoram mountains in northern Pakistan which have virtually no vegetation that would bind the surface. As a result, monsoon rains regularly induce landslides and avalanches. The

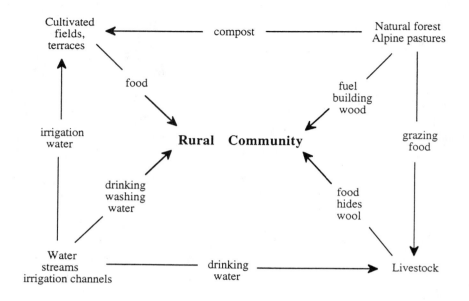

Figure 2. Land use, livestock management and water utilization by a mountain rural community (modified from Panth and Gautam, 1990).

1992 autumn rains in the Gilgit area and the 1997 summer rains in the Chitral area are recent examples of damage that the authors have personally experienced. Even at lower altitudes the removal of trees by village communities and the reduction of ground vegetation cover by grazing goats can destabilize slopes to such a degree that massive soil erosion develops. This soil erosion is one of the main effects of deforestation. The hillsides are denuded of vegetation and soil erosion follows. The soil is transported from higher levels and causes silting problems in rivers and dams. For example, the Tarbela Dam north of Islamabad in Pakistan is silting up to such a degree that the dam will be virtually useless by about 2030. Mountain agriculture can change very rapidly. Within the last six years (1992-1998) large numbers of poultry farms have been built in the Himalayan foothills adjacent to the road linking Islamabad and Murree in northern Pakistan (personal observations). The immediate impact is obvious: they are unsightly and smelly. The environmental impact has as yet to be assessed, although apparently much of the poultry farm waste enters local tributaries of the Indus. This will almost certainly affect rural populations that use the water downstream.

Rivers

River environments also contain a range of ecosystems and unique hydrological, geographical and geological features. Historically, rivers have been important centres where human civilizations have evolved and flourished. Examples are the ancient civilizations that developed along the Indus, the Ganges and Brahmaputra, the Nile, Euphrates and Tigris (Majumdar et al., 1967; Wheeler, 1968; Possehl, 1979; 1993; Shroder, 1993). Man has used river water in a number of ways (Snelgrove, 1967; Andrew, 1986; Meadows and Meadows, 1998a). These include irrigation and agriculture, transportation, fishing, generating electricity (hydroelectric power), domestic water supply, and the dumping of industrial and human waste.

Rivers serve as routes for travel and trade and also allow the development of high quality agricultural land along their banks (McDonald and Kay, 1988). This latter use can be extended hundreds of kilometers if canal irrigation is present. Rivers are particularly important for providing agricultural land in the Asiatic region as many of the large rivers are tropical or subtropical and continuous supplies of surface or ground water are vital.

The natural inputs and outputs of water and sediment to and from river systems are obvious. Water, dissolved materials, particulates and sediment are transported from mountains and hillsides. The sediment is progressively deposited on the river bed or on the surrounding plain by flooding. Overbank flooding during the monsoon season provides water that contains dissolved and particulate material which enriches the soil on flat flood plains. Part of the river flow also replenishes ground water which in many Asian countries is a vital source of water for domestic and agricultural purposes. The remaining water with its dissolved and fine particulate load then flows into the sea via estuaries and deltas. During this latter process much of the dissolved organic material is precipitated by the progressively increasing salt concentration as the freshwater mixes with sea water. The precipitated material forms the rich organic sediments characteristic of deltas and estuaries. This allows large populations of infaunal invertebrates to develop that then serve as food for indigenous and migrating birds. It also produces ideal conditions for the juvenile stages of many commercially important marine species of shrimp and fish.

These are the natural processes that have formed riverine, flood plain and deltaic environments over millennia. However, recent activities by man have shown that riverine systems are in a dynamic equilibrium between water and sediment inputs and outputs. Recent experiences of the effects of dams and barrages, and of excessive water withdrawal for irrigation and hydroelectricity have shown the dramatic effects that these activities can have on rivers and the surrounding countryside in Asian countries. Dams hold up water and sediment, canal systems and water channels export water from rivers, and routine heavy inputs of industrial and domestic pollutants destroy aquatic and wetland biodiversity. This is very obvious in the Indus Delta, where the reduction in freshwater input has resulted in degradation of mangrove swamp habitats and a consequent decline in fisheries.

Most of the large rivers in Asia are affected by man's activities of this sort. As a result, there have been many international and national commissions, inquiries, symposia and sets of recommendations. As an example, much of the water of the Indus River in Pakistan is lost to the huge canal system and to agricultural land (Tyres, 1978, quoting WAPDA data; Meadows and Pender, 1992). There are two major dams on its upper reaches, the Tarbela Dam near Islamabad and the Mangla Dam near Lahore. There are also a considerable number of barrages, link canals and major irrigation canals along its length (Figures 3, 4). The Water and Power Development Authority of the Government of Pakistan (WAPDA) have studied the distribution of river water between the Indus River and its canals and water courses, at the same time assessing the water being transferred to ground water and to agricultural land (Figure 5). Sixty six percent of the river flow is exported to the massive canal system. Of this about 50% enters the extensive system of water courses that irrigates agricultural land in the Punjab and in Sindh. There are also significant losses by evaporation and an input of 13% into ground water. The results of this transfer of water between the river and its tributaries, the canals, and agricultural land, is that only about 26% of the Indus' water reaches the deltaic coastal zone near Karachi. If projections into the 21st century are to be believed, this figure is likely to fall to below 15%.

The present impact of this on the coastal mangrove swamp areas in terms of biodiversity and artisanal fisheries is already extensive. An additional reduction of river flow by the introduction of more dams and the use of more water for agriculture and hydroelectricity during the early decades of the next century will pose very difficult

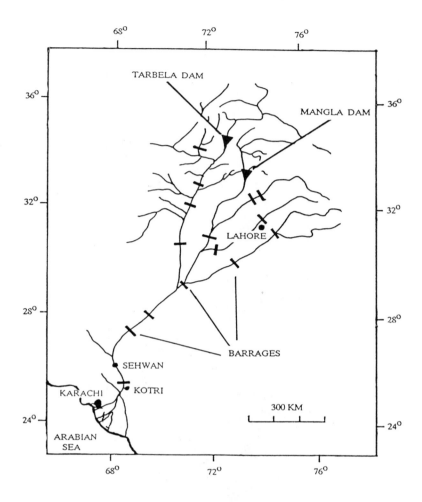

Figure 3. The river Indus and its tributaries with Tarbela Dam and Mangla Dam and a number of Barrages (from Milliman et al., 1984; Meadows and Pender, 1992).

environmental and political questions that will need to be addressed beforehand, and similar questions will have to be faced by other countries in the Asiatic region. The problem is an extremely difficult one. On one hand human populations using the river's resources need water. On the other hand many of the species and much of the ecosystem biodiversity in deltas and mangrove swamps depend on

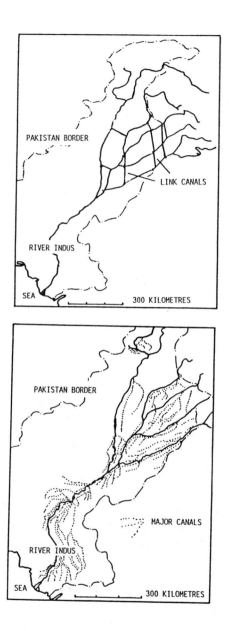

Figure 4. Link canals along the River Indus and its tributaries (top diagram); Major canals along the River Indus and its tributaries (bottom diagram) (Meadows et al., 1994).

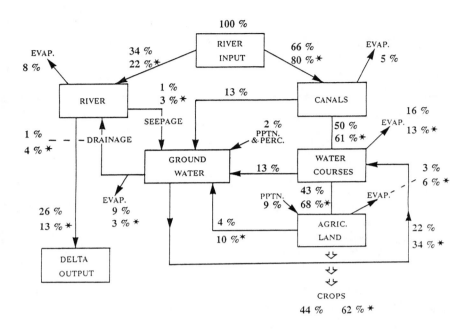

Figure 5. Different uses of Indus River water. The non-* percentages are inputs for the 1970s while the projected figures for the year 2000 are *, indicating considerably more water will be used in the canals, water courses, and agricultural land while the input to the delta will be halved (modified from Tyres, 1978, based on WAPDA, Government of Pakistan, plan).

significant inputs of freshwater, and of dissolved and organic particulate material. A balanced strategy is required to manage this important issue.

There are other problems with the over-use of river systems that are particularly prevalent in some Asian countries. These are desertification, salinization, sodification and waterlogging the surrounding flat alluvial land that is irrigated directly or indirectly by river water (Ellis and Mellor, 1995; Meadows et al., 1994). Desertification is a major problem in arid areas of mountain — river—coastal zone ecosystems in Asia. Desertification may be caused by both natural and man made processes, but the driving force appears to be a reduction of long term plant productivity and changes

Table 3. World's extent of dryland zones. P : PET = Precipitation : Potential evapotranspiration ratio

Type of Dryland Area	P : PET	Million hectare	Percent World Land
Dry - subhumid	0.45 - 0.70	1294.7	9.94
Semiarid	0.20 - 0.45	2305.3	17.72
Arid	0.05 - 0.20	1569.1	12.06
Total drylands susceptible to desertification		5169.1	39.72
Hyperarid (very desertlike, therefore, not susceptible to desertification)	<0.05	978.2	7.52

(Source: UNEP, 1992; Thomas, 1995; Watson et al., 1996)

in soil characteristics. This leads to the land becoming desert-like. Man made causes are over-cultivation, over-grazing, deforestation and inappropriate irrigation (Hellden, 1991; UNEP, 1992; Thomas, 1995; Watson et al., 1996). The first international recognition of the problem came with the United Nations Conference on Desertification held in Nairobi in 1977. The problem is caused by land degradation that occurs in susceptible drylands which may then develop into deserts if mismanaged. These drylands are technically defined as dry subhumid, semi-arid and arid drylands, and represent almost 40% of the total land area of the earth (Table 3).

The distinction between the three categories of susceptible drylands is based on the ratio of precipitation to potential evapotranspiration (P:PET) (UNEP, 1992; Watson *et al.*, 1996, pp178). Hyperarid drylands are not considered to be susceptible to desertification as they are already very desertlike. Asia has the highest susceptible dryland area of all the continents, 1671.8 million hectares. It also has the highest area that shows desertification, 370.4 million hectares (Table 4). The effects are very noticeable near many river systems in Asia. and the authors have observed many examples in the Sindh province of Pakistan, an area through which the Indus River flows as it approaches the coast of the Arabian Sea.

Table 4. Global extent of desertification in susceptible dryland areas. Units are millions of hectares.

Continent	Total susceptible dryland area (a)	Susceptible dryland area having		Total desertification (b+c=d)	Percentage desertification ((d/a)x100)
		Light and moderate desertification (b)	Strong and extreme desertification (c)		
Africa	1286.0	245.3	74.0	319.3	24.8
Asia	1671.8	326.7	43.7	370.4	22.2
Australasia	663.3	86.0	1.6	87.6	13.2
Europe	299.7	94.6	4.9	99.5	33.2
N. America	732.4	72.2	7.1	79.3	10.8
S. America	516.0	72.8	6.3	79.1	15.3
Totals	5169.2	897.6	137.6	1035.2	

(Source: UNEP, 1992; Thomas, 1995; Watson et al., 1996)

Salinization, in which salts accumulate in soil, occurs where the water table is close to the surface of the soil (White, 1987; Ellis and Mellor, 1995). It is caused by evaporation exceeding precipitation and is, hence, a common feature of semiarid and arid environments. Salinization is well known on flat flood plains like those of the Indus and on hillside slopes where the environment is dry. On flood plains, salinization occurs where the water table is close to the soil surface. This can be exacerbated by over-irrigation, and by leakage from poorly maintained canals and irrigation systems as in the Indus flood plain. It can also be caused by overgrazing, by deforestation, and by the replacement of deep rooting trees by shallow rooting grasses (Lavado and Taboada, 1987). The associated problem of soils becoming alkaline due to a dominance of sodium ions in the soil exchange complex (sodification), is more commonly associated with sloping hillsides on the edge of flood plains. Both salinization and the associated problem of sodification have detrimental effects on plant growth because the chemistry of soil nutrients is changed.

Waterlogging and salinization often occur together. As with salinization, waterlogging develops where the water table is close to the surface of the soil, often within one to three meters. In some areas the water table can even lie at the soil surface.

Table 5. Soil degradation caused by water erosion, wind erosion, salinization/sodification, waterlogging, and flooding (approximate percentage area). Comparison of deltaic, flood plain and mountain environments as illustrated by the five provinces of Pakistan (Sindh, Punjab, Balochistan, North West Frontier Province (NWPF), Northern Areas (NA).

	Flood plain and Deltaic (Sindh)	Flood Plain (Punjab)	Mainly Mountain (Balochistan)	Mainly Mountain (NWFP)	Mainly Mountain (NA)
Water erosion	4	19	76	92	99
Wind erosion	16	36	9	<0.1	<0.1
Salinization/ sodification	56	25	15	<0.1	<0.1
Waterlogging	12	5	<0.1	3	<0.1
Flooding	12	15	<0.1	5	<0.1

(Source: NWFP/IUCN, 1996)

Waterlogging can have a dramatic effect on plant productivity and the type of plant species that can flourish. In parts of the Indus Delta in Sindh, Pakistan, up to 12% of the land may be affected in this way (EUAD and IUCN, 1992; NWFP/IUCN, 1996). A comparison of the percentages of soil affected by salinization, sodification and waterlogging, together with water erosion and wind erosion shows the very significant differences between deltaic and flood plain areas on the one hand, and mountain areas on the other (Table 5) (NWFP/IUCN, 1996).

The Pakistan provinces of Sindh and the Punjab are made up of low lying deltaic and flood plain country surrounding the lower reaches of the Indus River. In these provinces, soil degradation is mainly caused by salinization, sodification, and wind erosion, with water erosion, waterlogging and flooding also having significant effects. The Pakistan provinces of Balochistan, the Northwest Frontier Province and the northern areas are largely mountainous, and include the upper reaches of the Indus River and many other smaller feeder rivers. In these provinces, soil degradation is mainly or almost entirely caused by water erosion.

The recent history of the management of the Indus River and its associated lands in Sindh is a classic example of how important salinization and waterlogging can become in deltaic areas and how difficult it is to find effective remedial measures. Snelgrove (1967) considers the problem of salinization and waterlogging in some detail in his book on the geohydrology of the Indus River. He lists a large number of recommendations from Pakistan's Water and Power Development Authority (WAPDA) that would reduce the problem considerably, and these recommendations are still relevant. However there is a lack of finance for many of the measures suggested by WAPDA. Part of the problem lies with the canal system. The original canal system constructed around it by the British in the 19th Century and later extended by the Pakistan Government is the largest in the world (Figure 4). However, with hindsight, inadequate subsurface drainage was installed during its construction. The present day effects of this are dramatic. An estimated 22 million tons of salt per year are carried by the Indus from the Himalayas as dissolved material. Much of this is deposited on the flood plains of the Indus valley in the Punjab and in Sindh province. The significant point is that only half of the yearly input reaches the Arabian Sea. The remainder of this salt, which represents about one ton for every hectare of irrigated land, becomes incorporated in the soil on evaporation of the water. It appears that many thousands of hectares of land are abandoned each year because of this salinization, to the extent that there is also a serious threat to one of the World Heritage Sites — Mohenjo Daro. The answer is to provide a route by which the water can drain from the ground so that it eventually enters the Arabian Sea with its load of dissolved material, including salt. Pakistan attempted this in the late 1980s by borrowing US$ 600 million to construct a left bank outfall drain to carry salt laden ground water to the Arabian Sea. However, this is not the whole story and may not be the best solution.

In the 1960s, the USA and WAPDA produced a report entitled "Land and Water Development in the Indus Plains of West Pakistan", which became known as the Revelle Report after its chairman Professor R. Revelle. The results of this report and the more recent Tyres Report of 1978 are still relevant and will serve as a model for similar situations into the 21st Century throughout Asia, wherever desertification, salinization and waterlogging are associated with the use of river water in agriculture. Both reports consider that the realization of the potential of the Indus River will be through agricultural development and modern technology. High crop yields could be achieved by the combined use of surface and ground water, and an increase in irrigation water is more important than more efficient use of water on farms. In high water table areas vertical drainage is feasible. Where the ground water consists of sweet water (water of very low or negligible salinity fit for human consumption), it could be used for irrigation of crops. Where the ground water contains appreciable quantities of salt, it could be exported by drainage to the ocean. Crops are usually underwatered at present, and this in itself increases soil salinity. Irrigation should be at a high enough level to allow evapotranspiration needs to be met in full with enough water remaining to be drained to the sea in order to carry away salt. The two reports also emphasized the need for more data on salt balances in the river system of the Indus, its associated irrigation systems, and the soil, in order that a net export of salt and water from the whole system can be guaranteed.

Coastal zones

Coastal zones in the Asian region are centres of large human populations and are also of global importance because of their very high levels of biodiversity. If one considers the coastal zone as including the 10 kilometers from low tide level both seaward and landward, it is obvious how significant the zone is for *Homo sapiens*, and this is especially so in Asia (Thompson and Tirmizi, 1995). There are many very large coastal cities with areas that are low lying and, therefore, prone to flooding, and some areas that are affected by seismic activity either directly by earthquakes or indirectly by tsunami. One only has to think of countries such as the Indian subcontinent, Indonesia, Japan, Malaysia, Papua New Guinea, the Philippines and Thailand. Many of the coastal cities and artisanal villages in these countries are under continuous threat. A visit to any of them shows the scale of the coastal environment and its present and future problems. For example, the commercially important habitats of 612,370 hectares of mangrove swamps on Pakistan's coastline are already under threat due to man's use of mangrove wood for timber, firewood and fodder. It has been estimated that this mangrove cover decreased from 43% to 23% during 1990 (IOC, 1994).

Coastal zones are currently under threat from a number of environmental and man-made pressures (CIRIA/CUR, 1991; Healy and Doody, 1995). Industrial and domestic pollution, habitat destruction by land reclamation, and over-exploitation of natural resources by towns and villages are all taking place at the present time. There is, therefore, an urgent need for new forms of resource utilization in the coastal zone, and in particular the sustainable development of coastal village community systems and the use of novel marine biotechnology (Zilinskas and Lundin, 1993; IOC, 1994). These should include sustainable development such as that being considered for the village of Kaka on the coast near Karachi, Pakistan (IOC, 1994). The fishing village of Kaka is the center of an experiment run by University, NGO, and governmental organizations in Sindh, Pakistan, whose objectives are to provide a model for the sustainable development of other coastal villages in the region. The experiment includes health care, education, and sustainable use of nearby terrestrial and marine ecosystems.

New approaches to coastal mariculture of seaweeds and inshore fish, perhaps using transgenic or triploid species, should also be developed (Zilinskas and Lundin, 1993). A triploid species of the oyster *Crassostrea gigas* is known that contains three rather than two sets of chromosomes and is sterile. It grows considerably faster than normal non-sterile diploid strains, and does not suffer from poor flavor associated with seasonal breeding that is a characteristic of the normal diploid species (Allen, 1988). Allen reports that 50% of the United States northwest's production of oysters comes from these triploid oysters.

Other areas that are being rapidly developed in the coastal zone include the highly successful plantations of a range of mangrove tree species in the Indus Delta, led by Mr. Tahir Qureshi working originally from the Sindh Forest Department and now from IUCN in Karachi, Pakistan (Qureshi, 1990). The work was started in 1986 as a five year research project funded by the Sindh Forest Department and UNDP/Unesco. Qureshi reported in his paper of 1990 that trials had been conducted on 17 species in 1986. These were followed by trials in 1987 involving four indigenous species of mangrove tree (*Aegiceras corniculatum*, *Avicennia marina*, *Rhizophora mucronata*, and *Ceriops tagal*) and one exotic species of mangrove tree from Bangkok (*Lumniztera racemosa*). Subsequent work extended the species list. Qureshi concluded that amongst the indigenous species, *Rhizophora mucronata* and *Avicennia marina* were the most promising. The field research has been considerably extended in the 1990s. The plantations visited by the authors in autumn 1997 are extensive, and consist of a number of nursery sites as well as mature stands of different mangrove tree species on the Sindh and

Balochistan coast of the Arabian Sea. This classic work, which involves active participation by local rural communities, will eventually lead to the revitalization of large areas of mangrove swamps not only on the Pakistan coast but in the coastal areas of other Asian countries.

With hindsight, however, the potential rise in sea level caused by global warming over the next 50 years may be the most important factor in coastal zone development. The difficulty with global warming is the lack of precision in predictions and the lack of agreement about how much of the potential warming is caused by man's activities as compared with natural long term perturbations in weather patterns. It is undoubtedly true that atmospheric carbon dioxide has been rising since the beginning of the industrial revolution in the early 19th Century. There is also a convincing body of evidence showing that other greenhouse gases such as methane, nitrous oxide, chlorofluorocarbons (CFCs) and hydrochlorofluorocarbons (HCFCs) are being added to the atmosphere by man's activities (Table 6). All of these gases will certainly add to global warming by acting as a blanket in the upper atmosphere It is particularly worrying that CFCs and HCFCs have effects that are two or more orders of magnitude greater than methane or carbon dioxide (Table 6 column 2).

The greenhouse gases reduce the amount of long wavelength radiation in the infrared that the earth radiates into space. Hence the temperature of the atmosphere increases. Predictions suggest an increase of 0.5°C to 2.5°C by 2030, which is likely to lead to a rise in sea level of between 10 cm and 20 cm. At first sight this does not appear to be a large increase, but calculations show that in very low lying areas such as on the east coast of parts of England and in the deltaic region of the Ganges and Brahmaputra, the effects are likely to be catastrophic. Floods in these coastal regions caused by high river flows and by abnormally high sea levels associated with low atmospheric pressures, already cause massive loss of life.

There may also be a change in weather patterns and increased frequency of severe storms. Even if worst case scenarios do not occur, it is undoubtedly true that within the last decade there have been significant changes in coastal weather. The southern parts of England have become much drier during summer, and the coastal zone of Pakistan especially in the Karachi area does not now have a winter rainy season. These changes are obvious even to the ordinary person, and are already having an impact on the economy of the

Table 6. Greenhouse gases: their relative effect, concentrations in the atmosphere, current rate of change and lifetime.

Greenhouse gases	Approximate relative greenhouse effect (GWP)	Current average atmospheric concentration (ppmv) (1992)	Current rate of change (% p.a.)	Atmospheric lifetime (years)
Carbon dioxide	1	355	1.8	50-200*
Methane	11	1.72	0.8	12
Nitrous oxide	270	0.31	0.25	120
CFC 11	3400	0.00026	4	50
CFC 12	7100	0.00045	4	-
HCFC 22	1600	0.0001	-	12

GWP: Global warming potential; ppmv: part per million by volume; * carbon dioxide does not have a single lifetime because the uptake rates differ for different sink processes. Global warming potential is the predicted reduction in the amount of heat radiated from the earth produced by a particular gas, in relation to the reduction produced by carbon dioxide over the same time period. It is only a very approximate measure of the relative effect of a greenhouse gas, as errors of up to 35% are recognized.
CFC:Chlorofluorocarbons; HCFC:Hydrochlorofluorocarbons
(Source: The Department of the Environment, n.d.; Houghton et al., 1996.)

areas affected. Conventional crops are suffering, and new cultivars are being assessed that can withstand drier, hotter conditions.

Coastal areas short of drinking water are being forced to introduce water rationing, and desalination plants. Such plants were once only seen in the more arid parts of the tropics in countries such as Libya and Saudi Arabia, but now are in operation in southern England. In Asia the main problem is often one of finance, as many of the countries under threat are underdeveloped. As a result, massive loans from such sources as the United States of America, Canada (CIDA), Great Britain (ODA—now DFID, the Department for International Development), UNESCO, and the World Bank are needed to initiate industrial activity. The loans almost always come with provisos. These provisos are usually needed to maintain some degree of fairness and lack of corruption, but sometimes they are of a political nature which may be less desirable.

Environmental Management and Community Involvement: Conclusions and Recommendations for Action in the 21st Century

One of the most important requirements for future work in the 21st Century on mountains, rivers, and the coastal zone is for the development of coherent multidisciplinary environmental programs of research and education involving an integrated and holistic approach. This will involve both urban and rural communities. There will also be a need for new information systems both for teaching and for field research. The multidisciplinary programs will require detailed cooperative plans to be formulated during the early stages of program development and will need careful coordination at a local, national, and international level.

A major problem which faces the developing countries is high population growth. "If current high rates of population growth and increasing rates of resource consumption do not drop, growing demand for resources will overwhelm even the best managed production systems" (WRI et al., 1992, pp. 49). For example, the energy consumption per capita is 1 gigajoule in a developing country with a high population growth like Bangladesh, 110 gigajoules in a developed country like Japan and 280 gigajoules in a highly developed country with a low population growth like the USA (WRI et al., 1992). In order to attain a sustainable environment, developing countries need to strike a balance between economic growth, efficient use of resources and the control of population growth. Developed nations on the other hand should reduce their over-use of energy and consumption of natural resources in order to promote and share global sustainability.

There is also an urgent need for integrated education and research involving conservation, management, protection, and restoration. The educational aspect of this will include the education of politicians and members of urban and rural communities. There will be a need for the development of new information systems for remote continuous monitoring of the environment. The integrated education and research programs should include: (1) biodiversity of endangered ecosystems, conservation of endangered habitats and species, and databases of flora and fauna; (2) catastrophic flood hazards and flood control; (3) deforestation and its effects on erosion and land usage; (4) desertification, salinization, and waterlogging; (5) slope stability on hillsides, landslides, and earthquakes; (6) environmental consequences of damming; (7) geological and biological history of river drainage systems; (8) impact of sea level

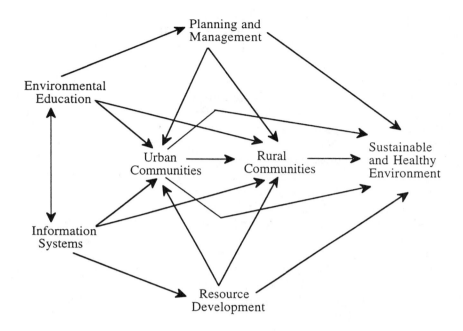

Figure 6. Environmental education, information systems and community integration.

rise; (9) mountain and river basin minerals, sediments, stratigraphy, tectonics, and formation; and (10) archaeology of ancient sites of civilization: early man including Stone Age sites (earliest evidence of man in Asia, ca 2,000,000 B.P.). These target areas require to be identified country-by-country. In this context there is an urgent need for the governments of developing Asian countries to act to conserve their environment by selecting conservation areas, nature reserves, heritage sites, sites of special scientific and commercial interest, and sites for tourist development. These regions will have to be environmentally assessed, controlled and monitored by governmental and non-governmental organizations and committees of national and international experts.

Finally, it is vital to develop public awareness of environmental issues and to include public participation in environmental programs, particularly in the developing nations of Asia (Figure 6). This should extend from the highest political level to members of the most remote rural communities. In the current decade of the information boom, particularly in developing countries, it is the urban population which has access to information systems. Rural communities are almost totally uninformed on this front. In order to bring the two major levels of the public to a stage of intercommunication we have to start educating rural communities (NWFP/IUCN, 1996). This in turn will help in poverty alleviation (Hussain, 1994). There is a strong argument for this taking precedence over the advancement of information systems which are only used by urban populations. The key issue is to bridge the gap between urban and rural communities, and develop an interacting environment for a sustainable nation. At a rural level modern information technology is often not available, either due to lack of resources or due to remoteness. In these communities, small-scale adult education programs should be introduced which use information technology to address matters such as health, farming and water use. Such programs will enhance the quality of life of isolated communities. At a popular level, environmental education and training in schools and universities, and media coverage of environmental matters can also provide a means of introducing the understanding of environmental problems and their management to local communities. This approach should emphasize the importance of human impact, resource and ecotourism potential, socioeconomic development, and cultural significance in promoting environmental values.

Acknowledgments

The authors are most grateful to the organizing committee of the Shiga University International Symposium 1997 for their invitation to present a paper, and for the excellent facilities and kind hospitality during the symposium. They would also like to thank the University of Glasgow for leave of absence to attend the symposium.

References

Ahmed, R. and A.M. Sheikh, Ed. 1994. *Geology in South Asia - I. Proceedings of the First South Asia Geological Congress, Islamabad, Pakisatan, Feb. 23-27, 1992.* Hydrocarbon Development Institute of Pakistan, Islamabad, Pakistan.

Allen, S.K. 1988. Triploid oysters ensure year-round supply. *Oceanus* 3 1: 58 - 63.

Andrew, W.P. 1986. *The Indus and its Provinces, Their Political and Commercial Importance.* Indus publications, Karachi, Pakistan, (originally published c. 1858).

Baquar, S.R. 1995. *Trees of Pakistan. Their Natural History, Characteristics and Utilization.* Royal Book Co., Karachi, Pakistan.

Bashir, E. and Israr-ud-Din, Ed. 1996. *Proceedings of the Second International Hindukush Cultural Conference.* Oxford University Press, Karachi, Pakistan.

Batterbury, S., T. Forsyth and K. Thomson. 1997. Environmental transformations in developing countries: hybrid research and democratic policy. *The Journal of Geography.* 163: 124-132.

Burrough, P.A. 1986. Principles of Geographical Information Systems for Land Resource Assessment. Clarendon Press, Oxford, UK.

Byerlee, D. and T. Husain. 1992. *Farming Systems of Pakistan: Diagnosing Priorities for Agricultural Research.* Vanguard Books, Lahore, Pakistan.

CIRIA/CUR. 1991. *Manual on the Use of Rock in Coastal and Shoreline Engineering.* Construction Industry Research and Information Association Special Publication 83, London, UK.

Dobby, E.H.G. 1962. *Monsoon Asia. Volume V of a Systematic Regional Geography.* 2nd edition. University of London Press, London, UK.

Ellis, S. and A. Mellor. 1995. *Soils and Environment.* Routledge, London, UK.

Embleton, C., Ed. 1972. *Glaciers and Glacial Erosion.* Macmillan, London, UK.

EUAD, GOP and IC. 1992. *The Pakistan National Conservation Strategy.* Karachi: Environmental and Urban Affairs Division, Government of Pakistan/Journalists' Resource Center for the Environment & International Union for the Conservation of Nature and Natural Resources, Karachi, Pkistan.

Gerrard, A.J. 1990. *Mountain Environments: An Examination of the Physical Geography of Mountains.* Belhaven Press, London, UK.

Groombridge, B., Ed. 1992. *Global Biodiversity.* Chapman & Hall, London, UK.

Hawtin, G.C. and N. Mateo. 1990. Mountain agriculture and crop genetic resources, In: *Mountain Agriculture and Crop Genetic Resources,* Riley, K.W., Mateo, N.M., Hawtin, G.C., and Yadav, R. Eds. Oxford & IBH Publishing Co., New Delhi, India, pp. I -7.

Healy, M.G. and J.P. Doody, Ed. 1995. *Directions in European Coastal Management.* Samara Publishing Ltd., Cardigan, UK.

Hellden, U. 1991. Desertification — time for an assessment. *Ambio.* 20: 372-383.

Houghton, J.T., L.G. Meira Filho, B.A. Callander, N. Harris, A. Kattenberg and K. Maskell, Ed. 1996. *Climate change 1995. Contribution of Working Group I to the Second Assessment Report of the Intergovernmental Panel on Climate Change.* Cambridge University Press, Cambridge, UK.

Hussain, A. 1994. *Poverty Alleviation in Pakistan.* Pakistan: Vanguard Books, Lahore, Pakistan.

IOC. 1994. *International Workshop on Integrated Coastal Zone Management, Karachi, Pakistan, 10-14 Oct. 1994.* IOC Workshop. Report no. 114. : UNESCO/Intergovernmental Oceanographic Commission, Paris, France.

IUCN and UNEP. 1986a. *Review of the Protected Areas System in the Indo-Malayan Realm.*: The World Conservation Union, Gland, Switzerland.

IUCN and UNEP. 1986b. *Review of the Protected Areas System in the Afrotropical Realm.*: The World Conservation Union, Gland, Switzerland.

IUCN 1990. *IUCN Directory of South Asian Protected Areas.* IUCN, The World Conservation Union, Gland, Switzerland and Cambridge, U.K.

Lavado, R. S. and M.A. Taboada. 1987. Soil salinization as an effect of grazing in a native grassland soil in the Flooding Pampa of Argentina. *Soil Use and Management.* 3(1): 143-148.

Majumdar, R.C., H.C. Raychaudhuri and K. Datta. 1967. *An Advanced History of India.* 3rd edition. St Martins Press, New York, USA.

McDonald, A.T. and D. Kay. 1988. *Water Resources: Issues and Strategies.* Longman, Harlow, Essex, UK.

Meadows, P.S., G. Pender and A. Meadows. 1994. The Indus River and Pakistan's economy: energy, environmental resources, and developmental policy. *Science, Technology & Development.* 12: 40-50.

Meadows, P. S. and G. Pender. 1992. The Indus river ecosystem and its use by man. International Conference on Protection and Development of the Nile and Other Major Rivers, Vol. 1/2, 4-9-1 to 4-9-15. Organized by the Nile Research Institute NRI, WRC and SNC/NHC/ECG.

Meadows, A. and P.S. Meadows, Ed. 1998a. *The Indus River: Biodiversity, Resources, Humankind. Proceedings of the Linnean Society of London Symposium.* Oxford University Press, Karachi, Pakistan (in press).

Meadows, P. S. and A. Meadows. 1998b. The environmental impact of the River Indus on the coastal and offshore zones of the Arabian Sea and the north west Indian Ocean, In:*The Indus River: Biodiversity, Resources, Humankind. Proceedings of the Linnean Society of London Symposium.* Oxford University Press, Karachi, Pakistan (in press).

Milliman, J.D., G.S. Qureshi and M.A.A. Beg. 1984. Sediment discharge from the Indus River to the ocean: past, present and future, In: *Marine Geology and Oceanography of Arabian Sea and Coastal Pakistan*, Haq, B.U., Milliman, J.D., Ed. Van Nostrand Reinhold Co, New York, USApp. 65-70.

Nasir, Y.J., R.A. Rafiq, and T.J. Roberts, Ed. 1995. *Wild Flowers of Pakistan.* Oxford University Press, Karachi, Pakistan.

NWFP/IUCN 1996. *Sarhad Provincial Conservation Strategy.* IC, Peshawar, Pakistan.

Panth, M.P. and J.C. Gautam. 1990. Mountain farming systems in Nepal, In: *Mountain Agriculture and Crop Genetic Resources*, Riley, K.W., Mateo, N.M., Hawtin, G.C., and Yadav, R. Eds. Oxford & IBH Publishing Co., New Delhi, India, pp 51 -68.

Possehl, G.L., Ed. 1979. *Ancient Cities of the Indus.* Vikes Publishing House Co. Pvt. Ltd., New Delhi, India.

Possehl, G.L., Ed. 1993. *Harappan Civilization.* 2nd edition. Oxford and IBH Publishing Co. Pvt. Ltd., New Delhi, India.

Price, M.F. and D.I. Heywood, Ed. 1994. *Mountain Environments and Geographic Information Systems.* Taylor and Francis, London, UK.

Qureshi, M.T. 1990. Mangrove ecosystems. *UNDP/UNESCO Regional mangroves project RAS/86/120, Occasional paper no. 4.* UNDP/UNESCO, Paris, France.

Rengers, N., R. Soeters and C.J. Van Westen. 1992. Remote sensing and GIS applied to mountain hazard mapping. *Episodes.* 15: 36-45.

Riley: K.W., N.M. Mateo, G.C. Hawtin and R. Yadav, Ed. 1990. *Mountain Agriculture and Crop Genetic Resources.* Oxford & IBH Publishing Co., New Delhi, India.

Roberts, T.J. 1991. *The Birds of Pakistan Volume 1: Regional Studies and Non-Passeriformes.* Oxford University Press, Karachi, Pakistan.

Roberts, T.J. 1992. The *Birds of Pakistan Volume 2: Passeriformes Pittas to Buntings.* Oxford University Press, Karachi, Pakistan.

Roberts, T.J. 1997. The *Mammals of Pakistan.* 2nd edition. Oxford University Press, Karachi, Pakistan.

Shroder, J.F. Jr, Ed. 1993. *Himalaya to the Sea.* Routledge, London, UK.

Snelgrove, A.K. 1967. *Geohydrology of the Indus River West Pakistan.* Sind University Press, Hyderabad, Pakistan.

Soeters, R., N. Rengers and C.J. Van Westen 1991. Remote sensing and geographical information systems as applied to mountain hazard analysis and environmental monitoring, In: *Proceedings, 8th Thematic Conference on Geologic Remote Sensing, Denver,* Co, USA, vol. 2: 1389-1402, Ann Arbor, Michigan, USA.

The Department of the Environment. n.d. *Global Climate Change.* London, UK.

Thomas, D.S.G. 1995. Desertification, causes and processes, in *Encyclopaedia of Environmental Biology,* vol. I: 463-473. Academic Press, London, UK.

Thompson, M.-F. and N.M. Tirmizi, Ed. 1995.*The Arabian Sea. Living Marine Resources and the Environment.* Vanguard Books, Lahore, Pakistan.

Tyres, R., Ed. 1978. *Report of the Indus Basin Research Assessment Group. Research issues affecting agricultural development policy.* Planning Commission, Government of Pakistan, Islamabad, Pakistan.

UNEP 1992. World *Atlas of Desertification.* Arnold, London, UK.

Watts, R.T., M.C. Zinyowera and R.H. Moss, Ed. 1996. *Climate Change 1995. Impacts, Adaptations, and Mitigation of Climate Change: Scientific-Technical Analyses.*: Cambridge University Press, Cambridge, UK.

Wheeler, M. 1968. The *Indus Civilization.* 3rd edition. Book Club Associates by arrangement with Cambridge University Press, London, UK.

White, R.E. 1987 *Introduction to the Principles and Practice of Soil Science.* Blackwell, Oxford, UK.

WRI, IUCN and UNEP 1992 *Global Biodiversity Strategy. Guidelines for Action to Save, Study, and Use Earth's Biotic Wealth Sustainably and Equitably.* World Resources Institute, United Nations Environment Program, Food and Agriculture Organization, United Nations Educational, Scientific and Cultural Organziation

Zilinskas, R.A. and Lundin, C.G. 1993 *Marine Biotechnology and Developing Countries.* World Bank Discussion Paper 210, World Bank, Washington DC, USA.

Chapter 5

Environmental Problems and Public Awareness in the Republic of Korea

Jung Wk Kim, Kye Won Lee and Gyu Ho Jung, School of Environmental Studies, Seoul National University, Shinlim-dong, Gwarak, Seoul, 151-742, Korea.

The State of the Environment in the Republic of Korea

The present environmental problems have emerged in the Republic of Korea since the 1960s. The government launched the first 5-year economic development plan in 1962. Their aim was to industrialize Korea in the manner of Western countries. The government tried to import industries without any regard for the environment. Even mentioning environmental problems was regarded as "national treachery." The industrialized Western countries experienced serious environmental problems especially during the 1950s and 1960s. The industrialization in South Korea at that time offered a good opportunity for Western countries to dispose of their polluting industries.

The Office of the Environment (the former Ministry of the Environment) was established in 1980. Before that there had not been any effective means of environmental control. It may be better said that the government encouraged environmental pollution rather than preventing it. There used to be a "Pollution Prevention Law," but the law was so loose that industries could hardly violate the law even though they didn't operate any treatment facilities, and even when they violated the law, none of them were punished for it. The industrialized countries achieved noticeable improvements in their environment since the 1970s, while South Korea underwent rapid deterioration.

The industrialized Western countries have improved their energy efficiency considerably. The average energy consumption per GNP in OECD countries improved from 0.52 TOE (tons of oil equivalent)/ 1000 US$ in 1973 to 0.40 TOE/1000 US$ in 1989. The United States of America improved from 0.57 TOE/1000 US$ in 1973 to 0.43 TOE/1000 US$ in 1989, and Japan from 0.38 TOE/1000 US$ in 1973 to 0.17 TOE/1000 US$ in 1989. However, that in South Korea remained almost the same during that time: 0.60

TOE/1000 US$ in 1973 to 0.58 TOE/1000 US$ in 1989, which is one of the world's largest (KEEI, 1994: 44).

The industrialized countries have also succeeded in reducing pollution emission remarkably since the 1970s. For example, Japan emitted 4.97 million tons of sulfur dioxide in 1970, but the emission was reduced to 1.26 million tons in 1980 and 0.876 million tons in 1989 (OECD, 1993: 17-19). However, the emissions in South Korea increased: from 1.33 million tons in 1978 to 1.61 million tons in 1990, which well surpassed that in Japan. The size of the South Korean economy is only 1/12 of that of Japan, the population 1/3, and the area 1/4, but the sulfur dioxide emission is twice that of Japan. The annual growth rates of industrial waste water and industrial solid wastes were both 13% between 1980 and 1990, which surpassed the economic growth rate during that time, 9.6% (MoE, 1982-1996). Only in recent years has there appeared to be some signs of improvement: the sulfur dioxide emissions began to decrease from 1991 and the municipal solid wastes from 1992, but most of other pollutants are still increasing.

As a result, the quality of the environment has been deteriorating continually in South Korea. The annual average sulfur dioxide level at Shinsul-dong in Seoul reached 0.183 ppm in 1987 (Kim, 1990a), while the level in Western countries usually is below 0.01 ppm. Since then, the level dropped rapidly with Seoul reporting 0.017 ppm in 1995. However, the photochemical smog resulting mainly from increased automobiles has become a serious air pollution problem in major cities. Most people do not trust the public water supply as safe. According to polls, more than 95% of the Korean population do not drink water directly from the faucet (MoE, 1995). They either boil it, filter it with commercial purifying devices, or buy mineral water. The 10 million inhabitants of the Nakdong River Basin especially do not trust the tap water because the pollution of the river can apparently be detected by odor or taste. There have been several episodes involving spills of chemicals such as phenol and chlorobenzenes in the Nakdong River in recent years. The disposal of solid wastes has been one of the major sources of complaints from citizens also. Citizens near the incineration plants and landfill sites are staging vigorous protests either because of damage already inflicted on them or from fear of more in the future.

The future prospects of the Korean environment do not seem very bright if the South Korean economic development plan for the 21st Century is considered relative to the regional environmental problems in northeast Asia.

The energy demand in South Korea is expected to grow very fast to support the government's economic development plan. According to estimates by the Korea Energy Economics Institute, the energy demand in 1990 was 93.2 million TOE, but it is expected to grow to 178 million TOE in 2000, to 253 million TOE in 2010, and to 392 million TOE in 2030, about a four-fold increase from 1990 (KEEI, 1994: 46). This energy demand is expected to exert a direct and significant impact on the environment.

Korea also will be strongly affected by the trans-boundary environmental problems in the northeast Asian region. Due to the westerly wind prevailing in this region, air pollution in China can easily be transported to Korea. The amount of sulfur dioxide generated in China was estimated to be 17.5 million tons in 1990, about 11 times that in South Korea. It is estimated that about 1 or 2% of the sulfur dioxide generated in China is deposited in the Korean Peninsula either in wet or dry forms (Kim, 1993). If the air pollution emission in China increases in line with economic growth, which is the fastest in the world, it is expected to cause a serious problem in Korea. If China's economy grows at 10% a year, it will increase 20-fold in 30 years. If energy demand is assumed to grow at the economic growth rate, then the sulfur dioxide emission will be roughly 200 times that in South Korea. In addition, the Yellow Sea surrounded by China and the Korean Peninsula also suffers from serious problems. Currently, South Korea is discharging 460,000 tons of BOD per year into the Yellow Sea. When China stops using manure as fertilizer and discharges it to the sea as South Korea does, the sea will become badly contaminated with sewage and become eutrophic.

The problems underlying South Korea's environmental policy may be delineated as follows:

First, the wasteful use of energy has been encouraged by the government's cheap energy policy. Energy in Korea can be cheap because there has been insufficient investment in cutting down the pollution caused by energy use. Industries were not motivated to save energy because it was cheap. The social costs of the environmental pollution resulting from using cheap energy have also been great, which, in turn, requires the use of more energy.

Second, the waste of material resources has also been encouraged. Solid wastes are clear evidence of wasted materials. The per-capita solid wastes generated in South Korea exceed those in most of the industrialized countries. Disposable products, which are destined to become waste, flourish and there is no properly established recycling system. Moreover, the costs of disposing of disposable products and the damage they do to the environment are not reflected in their prices.

Third, the South Korean environmental policy is encouraging environmental pollution rather than discouraging it. The environmental standards have been loose compared with those in industrialized countries. Besides, even the loose standards are not well observed because the penalties for violating environmental laws have been regarded as cheaper than complying with them. Therefore, industries would rather risk being caught violating the laws than trying to meet the standards. The development of projects which create environmental burdens, such as industrial and resort development, has been faster than that of environmental projects such as waste water treatment. The fact that pollution emissions have grown faster than the economy as a whole is clear evidence of the effects of government policy.

Last, valuable natural ecosystems are being destroyed. These include the important forests and wetlands. Obtaining a permit to clear forest to construct a golf course or resort guarantees a fortune while preserving the forest means just paying taxes, which in turn encourages the destruction of forests. Because of the policy for popularizing golf launched by former president Rho Tae Woo in 1988, the number of golf courses increased from 21 to about 200, and many of these have been constructed on the site of cleared forests. The area of golf courses reached 20,292 ha in 1991, almost equal to the total industrial area in South Korea (Kim, 1991). The wetlands on the West Coast have also continued to disappear. The reclamation of these wetlands is considered to be a big economic loss at the national level. Nevertheless, the pressure for reclamation is great because the speculators earn high profits developing the reclaimed lands.

Environmental Episodes and Public Participation

Until South Korea was democratized in 1987, the mass media used to be strictly censored to filter out environmental news, and the public was not allowed to get involved in environmental movements. There had been severe environmental damage around the industrial complexes, but the public had to endure their sufferings because of the harsh dictatorship during that time.

After democratization in 1987, numerous environmental groups sprang up, and they have been leading participants in environmental protection. Existing NGOs such as religious groups and consumer organizations have expanded their activities to include the environment. Their activities initially focused on assisting pollution victims and investigating pollution damage in industrial areas, but later they diversified into environmental monitoring, education, campaigns, evaluating government policies, and assessing alternative policies, etc.

As the environmental damage in South Korea had been very severe, public participation also was violent in the beginning. Some cases of environmental campaigns in which citizens participated actively may be mentioned.

The earliest environmental problems observed in South Korea since industrialization were the damage inflicted on farmers and fishermen around areas like the Ulsan/Onsan and Yeochun industrial complexes. In 1985, the government finally decided to relocate about 37,000 inhabitants from the Ulsan/Onsan industrial complexes for fear of health problems as well as damage to agricultural crops and fisheries (Kim, 1990b). The victims staged violent demonstrations demanding compensation. Even though the public was forced to stay away from the dispute under the dictatorship at the time, this episode helped awake public awareness of the environment.

The environmental conflicts, which formerly used to be limited to industrial areas, began to expand into areas surrounding metropolitan cities or industrial estates and even into the rural areas as waste disposal sites or other environmentally damaging facilities were built there. Until that time, there had not been any reliable waste treatment facilities. In 1990, the residents in Bansong-dong, Pusan City, opposed the construction of an industrial waste disposal plant, blocked the road to the site, and staged a violent demonstration which led to the cancellation of the project as well as to the arrest of the protest leaders. In the same year, there was a violent riot in Anmyun Island, a remote island near the West Coast, opposing the government plan to build a nuclear waste disposal site. Government buildings and property were set on fire and there were many citizen casualties. This project also had to be canceled.

There also were movements against golf-course development. Golf courses in South Korea are usually constructed on steep hills, frequently on the site of former forests or on arable land. Golf courses built upstream of agricultural villages cause damage by reducing the flow of water and polluting it. These movements were usually not successful, even though they were just as violent as other movements, possibly because the numbers of residents participating in a movement were insufficient to cover the area of a large golf course. These people later formed a coalition and began a nation-wide movement.

In Korean society, like other oriental cultures, individual citizens are usually supposed to put up with inconveniences inflicted on them for the good of society. In such an atmosphere, the complaints of a minority of people suffering from environmental pollution in limited areas are easily overlooked. This tradition may be very dangerous when it comes to environmental issues. When an environmental problem becomes the problem of the majority, it may be too late to do anything about it. A good example is that of the pollution of the Nakdong River. It was reported that Doosan Electronics secretly discharged 325 tons of phenol waste into the river between November, 1990 and March, 1991. This practice ended as a result of a big spill of phenol in March, 1991. This caused tremendous turmoil among over 10 million inhabitants along the river basin, followed by complaints of diarrhea, sore throats, skin irritation, miscarriages, and other medical problems. The Nakdong River has experienced similar spill episodes almost every year since. Thirty-four environmental groups formed a special committee to fight against these environmental disasters, and their main target was the Doosan Group. After this ordeal, the Doosan Group has emerged as one of the most environmentally-friendly businesses in South Korea.

Citizens' concerns about the environment began with local environmental problems arising from certain development projects, but their concerns were gradually extended to include nation-wide, regional, and global environmental problems. In addition to the existing organizations, religious groups and local communities also began to show their concern about the environment. The members of these groups, along with other citizens, have been armed with environmental knowledge through various environmental education programs provided by academics and experts. Numerous environmental groups have emerged, and they have become very influential in leading public opinion. The mass media, in cooperation with these environmental NGOs, have played an important role in building up public awareness about the environment.

Public Environmental Awareness
Awareness of the environment

The Ministry of Environment has conducted or sponsored five survey studies on public awareness about the environment since 1982 (MoE, 1995; CRI, 1990; KETRI [Korea Environmental Technology

Table 1. The most dangerous threat to the human future.

Type of Threat	1982 %	1995 %
Environmental pollution	14.4 (3)	57.1 (1)
Population explosion/resource depletion	28.8 (2)	16.5 (2)
Natural disaster	9.8 (4)	10.8 (3)
World war (or nuclear war)	32.8 (1)	8.5 (4)
Moral degradation	4.5 (6)	6.9 (5)
Don't know	9.6 (5)	0.2 (6)

Source: MoE, 1982: 247-54; MoE, 1995: 204-09.

Research Institute], 1996). The results clearly show that awareness has increased very rapidly in recent years.

In 1982, nuclear war was ranked as the most dangerous threat to the human future by 32.8% of respondents because of Cold War tensions, while environmental pollution ranked third with 14.4%. But in 1995 the order was reversed: environmental pollution was now regarded as the most dangerous threat with 57.1% of responses (see Table 1).

The majority of people were pessimistic and were becoming more pessimistic about the future prospects for the environment. In 1982, 67.3% were pessimistic about the future environment, but in 1995, 75.9% were pessimistic (see Table 2).

People were getting more pessimistic about the environment, possibly because they thought the problems had been getting worse over the previous years. In 1982, 71.4% of the respondents answered that the environment was seriously damaged, but the proportion increased to 89.8% in 1996 (see Table 3).

Table 2. Prospects for the future environment.

Opinion	1982 %	1995 %
Environment will get worse	67.3	75.9
Environment will remain the same	18.2	16.7
Environment will be improved	9.2	7.4
Don't know	5.5	*

*This item was not included in the questionnaire.

Source: MoE, 1982: 245-47; MoE, 1995: 200-01.

Table 3. Opinions about the state of environmental pollution

Opinion	1982 %	1990 %	1996 %
Seriously polluted	71.4	77.2	89.8
So-so	22.4	19.8	*
Not serious	3.2	3.0	9.8
Don't know	3.0	*	0.4

* These items were not included in the questionnaires.
Source: MoE, 1982: 91-100; CRI, 1990: 33-39; KETRI, 1996: 4.

Table 4. Concern about the environment

	1990 %	9951 %
Concerned	81.0	82.4
Not concerned	19.1	17.6

Source: CRI, 1990: 28; MoE, 1995: 55-56.

Table 5. Public opinion of priorities in governmental policies

Year	Results
1982	(1) education (17.0%), (2) defense (13.5%), (3) food (11.7%), (4) prices (10.6%), (5) unemployment (9.5%), (6) consumer protection (7.2%), (7) environment (5.7%), (8) housing (5.6%)
1987	(1) income distribution (25.8%), (2) welfare (22%), (3) environment (17.1%), (4) education (13.9%), (5) traffic (6.8%), (6) housing (5.9%), (7) employment (5.6%)
1990	(1) security (23.7%), (2) environment (20.7%), (3) traffic/housing (20.6%), (4) low-income class (14.6%), (5) economic growth (10.3%), (6) education (9.7%)
1996	(1) environment (33.3%), (2) defense (16.4%), (3) traffic/construction (15.8%), (4) security (14.5%), (5) diplomacy (11.2%), (6) information / media (7.7%)

Source: MoE, 1982: 225-230; MoE, 1987: 175-181; CRI, 1990: 79-80; KETRI, 1996: 8.

The majority of the people said they were concerned about the environment. Since 1990, more than 80% of the respondents have answered that they were concerned (see Table 4).

The public thought the environment should be given a low priority among government policies until the early 1980s. In 1982 the issue ranked seventh, but in 1987 it was third, in 1990 it was second, and in 1996 it rose to the top, even ahead of defense (see Table 5).

To the question on the priority of policies for the 21st century, the environment earned the most votes (44.8%), followed by security (19.1%), information/media (16.9%), diplomacy (10.9%), defense (8.6%) and construction/traffic (7.4%) (KETRI, 1996: 8).

Evaluation of the government's policy

The Korean people consider tap water as a good indicator of the quality of the environment. Several episodes of drinking water pollution have led the people to distrust the government deeply. In response to the question on the progress of tap water quality in 1995, a majority answered that it was the same or worse: 39.3% answered that it was the same, 33.5% said it was worse, 8.4% said it had improved, and 18.7% said they didn't know (MoE, 1995: 68-69). The percentage of people who use tap water as drinking water has decreased, while the use of commercial mineral waters and mountain spring water has increased rapidly. It is notable that about 30% of the population relied on spring water in 1995: this is usually procured by the users themselves climbing mountains or commuting to parks. Less than 5% of the population trusted tap water to be safe for drinking (see Table 6).

The people seemed to link the quality of tap water directly to the government's performance. As the distrust of tap water grew, the evaluation of the government's environmental policy also deteriorated. In 1982, 68.3% of the population evaluated the government positively, but in 1996, only 38.2% said the government's efforts were satisfactory (see Table 7). The wording of the questions and answers differ slightly from year to year except in 1982 and 1987, so that a direct comparison between years is not completely possible.

Along with the poor evaluation of the government's performance, it turned out that people did not trust the government's environmental data either. For example, in 1996, 56.1% of the people surveyed answered that they did not trust this information (see Table 8).

To the question of who works the hardest for the environment, in 1990 the NGOs earned 51.2% of the vote, followed by the mass media (26.2%), the people (18.4%), and businesses (2.2%). The government ranked the last with only 0.8% of the vote (CRI, 1990: 59). And in a survey in 1996, 75.9% of the respondents answered they trusted environmental NGOs more than the government, while

Table 6. Source of drinking water

Drinking Water Source	1987 %	1990 %	1995 %
Boiled tap water	48.3	55.9	47.9
Boiled tap water/tap water	32.1	*	*
Tap water	6.7	*	2.8
Purified tap water	3.7	6.7	5.4
Ground water	6.3	18.4	9.8
Commercial mineral water	2.7	2.4	10.7
Mountain spring water	*	16.6	29.3

* These answers were not included in the questionnaires
Source: MoE, 1987: 88-90; CRI, 1990: 91; MoE, 1995: 63-66.

Table 7. Evaluation of the government's efforts

Evaluation	1982 %	1987 %	1990 %	1996 %
Very satisfactory	8.6	15.5	2.4	*
Satisfactory	59.7	41.7	22.0	38.2
So-so	*	*	30.7	*
Unsatisfactory	18.6	23.3	35.6	61.7
Very unsatisfactory	5.7	10.3	9.3	*
Don't know	7.4	10.6	*	0.1

*These items were not included in the questionnaires.
Source: MoE, 1982: 204-06; MoE, 1987: 163-65; CRI, 1990: 58; KETRI, 1996: 48.

Table 8. Opinions of reliability of government's environmental data

Response	1990 %	1995 %	1996 %
Trust very much	*	5.7	*
Trust a little	23.6	23.0	43.8
So-so	42.3	44.1	*
Do not trust much	33.8	22.8	56.1
Do not trust at all	*	5.7	*

* These answers were not included in the questionnaires
Source: CRI, 1990: 62; MoE, 1995: 131-34; KETRI, 1996: 49.

23.8% trusted the government more than NGOs. This clearly shows that the people's awareness of environmental issues is far ahead of that of the government (KETRI, 1996: 49).

Citizens' views and participation

The environmental investment in South Korea has been considerably lower than that of other OECD members. The South Korean government's investment ranged between 0.06 and 0.25% of the GNP (Ministry of Environment, 1982-1996) while in the OECD countries it ranged between 0.5 and 1% (OECD, 1993: 294-301). The government and industries fear that investment in the environment will not benefit the economy. But the 1996 surveys show that most people (86.6%) think environmental conservation is in fact beneficial for the economy (see Table 9). In the 1982 survey, people were slightly more favorable to economic development than to environmental conservation: 14.3% of respondents insisted the economy should be developed in spite of the sacrifice of the environment, 69.9% answered that the economy should be developed without sacrificing the environment and only 6.6% said economic development should be controlled to conserve the environment

Table 9. The role of environmental conservation in the economy.

Response	1990 %	1995 %	1996 %
Promotes economy	73.6	75.0	86.6
Half and half	*	20.6	*
Hinders economy	15.0	4.4	13.4
Don't know	11.4	*	*

* These answers were not included in the questionnaires
Source: CRI, 1990: 29; MoE, 1995: 110-13; KETRI, 1996: 6.

(MoE, 1982: 196-202). But in 1996, 85.2% answered that the environment was more important than the economy, and 82.3% answered that they were willing to share the cost for solving environmental problems (KETRI, 1996: 6).

While the citizens wanted the government to put more money and effort into the environment, they did not participate very actively in actual environmental movements. In 1990, 58.3% said they were willing to participate; 12.1% were against it; and the rest, 29.7% were in between (CRI, 1990: 70). But in 1995, 47.8% were willing to participate (a 10.5% drop from 5 years before), 18.4% were against it (a 6.3% increase), and 33.8% were in between. The percentage of those actually participating in the environmental activities turned out to be only 13.2% in 1995 (MoE, 1995: 60-62). The wide gap between those willing to participate and those doing so may mean that Korean society has more potential to mobilize citizens in environmental activities in the future.

To the question on the motivation for environmental concern in the 1990 survey, 43.2% said they had become interested in environmental problems through personal experiences, and 39.6% said that they had become interested through the mass media (CRI, 1990: 75). Of the various media, in the 1995 survey television turned out to be the biggest source of environmental information that citizens rely on (75.4%), followed by newspapers (18.4%) and magazines (3.1%) (MoE, 1995: 57-58).

Table 10. The number of environmental editorials in newspapers.

Year	'87	'88	'89	'90	'91	'93	'94	'95	'96
Chosun	8	13	15	16	23	27	20	43	44
Dong-A	9	15	9	11	13	22	32	18	21

Environmental reports in the mass media

The mass media play an important role in building citizens' awareness of the environment, while at the same time they also reflect citizens' thoughts on environmental issues because most citizens have access to environmental information through the mass media which report only the information citizens are interested in. Newspaper editorials especially are believed to play a very important role in leading public opinion. In the light of this view, the editorials of the two big newspapers in South Korea, Chosun Ilbo and Dong-A Ilbo, were analyzed.

The number of environmental editorials increased significantly during the last ten years: from 8 in 1987 to 44 in 1996 in Chosun, and from 9 in 1987 to 21 in 1996 in Dong-A (see Table 10).

The most common editorial issue was water pollution, 23.8% of the total during the four years from 1993 to 1996, followed respectively by the ecosystem (15.4%), nuclear issues (13.7%), solid wastes (12.8%), air pollution (6.6%) and others (27.8%). The environmental editorials usually appeared after big environmental events. According to the public opinion survey conducted in 1995, citizens gave first priority to the water pollution (33.8% of the responses), which probably was a reflection of the number of newspaper reports and editorials.

The party most frequently blamed for the problems in the editorials was the government (45.9% of the total between 1993 and 1996), followed by the people (15.4%), manufacturing businesses (14.7%), foreign countries (10.2%), and others (13.9%). And as for the party expected to clean up the mess, the government was most frequently cited (55.5%), followed by the people (19.3%), manufacturers (7.5%), foreign countries (5.0%), and others (12.8%).

Conclusions

Public awareness of the environment could be built up in the Republic of Korea only after the political system was democratized. After experiencing serious environmental disasters, the public has become well aware of the importance of conserving the environment. Public awareness has changed quickly, but the government's policy and the state of the environment have not been so responsive to it. The basic economic development plans were formulated during the dictatorial period and the public was only allowed to evaluate the environmental consequences after the plans were in effect. The government's decision-making processes have not been democratized at the same speed as the rest of society. For example, the two biggest developmental projects in Korea, Inchon International Airport and the Seoul-Pusan Bullet Train, were planned without opening up the decision-making process to the public, even though they were planned after the nation was democratized. It has been suggested that some institutional arrangements be established to reflect more effectively public opinion of the government's environmental policies by allowing the public to participate in the decision making process from the early stages.

Another important finding is that the public is willing to participate in environmental activities, but that their actual participation rate is very low. The Korean people have proved their willingness to help by faithfully observing newly proposed environmental protection systems like a separate collection system for recyclable wastes or food wastes, a volume-rate-collection system in which citizens are required to put wastes in prescribed bags, and others. But the new systems have not been working properly in many cases because they have not been properly set up by the government. For example, citizens collect wastes for recycling, but some of them are mixed again and simply dumped into landfills because there are not proper facilities to process the material. In addition, local communities are not well organized to make use the citizens' willingness and capabilities and interests either, probably because of the short history of local autonomy. South Korea re-started local autonomy only in 1991 after about thirty years of dictatorship. So it has been suggested that proper facilities and systems be established and local communities be organized autonomously, so as to mobilize citizens to participate actively in environmental activities and contribute to the betterment of the community.

References

CRI. 1990. *Public Opinion Survey on Environmental Conservation.* (Continental Research Institute), Ministry of Environment, Korea.

Kim, J.W. 1990a. Environmental pollution levels with indices in the Republic of Korea. *Shin Dong-A*, February: 498-513.

Kim, J.W. 1990b. Environmental aspects of transnational corporation activities in pollution-intensive industries in the Republic of Korea: a case study of the Ulsan/Onsan industrial complexes. *Environmental Aspects of Transnational Corporation Activities in Selected Asian and Pacific Developing Countries.* ESCAP (Economic and Social Commission for Asia and the Pacific/UNCTC (United Nations Center for Transnational Cooperation) Publication Series B, No. 15, United Nations, NY, USA, pp 276-319.

Kim, T.D. 1991. *The Socio-Economic Implications of Golf Course Development and Solutions.* Public Hearing sponsored by Citizens' Coalition for Economic Justice, Seoul, Korea.

Kim, T.Y. 1993. *A Study on the Effects of Air Pollution in China on the Korean Peninsula.* Master's Thesis, Graduate School of Environmental Studies, Seoul National University, Seoul, Korea.

KETRI. 1994. *A Mid- to Long-Term Policy for Strengthening Energy Demand Management.* (Korea Energy Economics Institute) KEEI Report 94-07, Euiwang, Korea.

KETRI (Korea Energy Economics Institute). 1996. *Public Awareness on Environmental Issues.* Korea Environmental Technology Research Institute, Seoul, Korea.

MoE (Ministry of Environment). 1982-1996. *Environmental White Book.* Ministry of the Environment, Seoul, Korea.

MoE (Ministry of Environment). 1982. *Public Opinion Survey Report on Environmental Conservation*, Ministry of Environment, Seoul, Korea.

MoE (Ministry of Environment). 1987. *Public Opinion Survey Report on Environmental Conservation*, Ministry of Environment, Seoul, Korea.

MoE (Ministry of Environment). 1995. *Public Opinion Survey Report on Environmental Conservation*, Ministry of Environment, Seoul, Korea.

OECD (Organization for Economic Cooperation and Development). 1993. OECD *Environmental Data Compendium* 1993. Organization for Economic Cooperation and Development, Paris, France.

Chapter 6

Economic Development and the Environment in Thailand: The Current Situation and the Role of Environmental Education and NGOs

Supachit Manopimoke, Faculty of Economics, Thammasat University, Bangkok 10200, Thailand.

Introduction

Prior to the current economic recession, Thailand had experienced a remarkable economic growth for over three decades. The growth rate of the country's gross domestic product (GDP) has been one of the highest and steadiest of the developing nations (Table 1). In fact, its GDP has been high and relatively stable since the early 1960s when the first economic development plan was implemented (Table 2). During 1960 to 1980, the average rate of real GDP growth was above 7% per year. The growth in GDP during 1987-1990 was even more remarkable, averaging over 10% per annum. In 1991, real GDP, estimated at US$ 28,648 million by the National Economic and Social Development Board (NESDB, 1991) was almost three times that experienced in 1980.

Table 1. Comparison of growth rates (in real GDP) of Asian countries

Country	1987	1988	1989	1990	Average
Thailand	9.5	13.2	12.0	10.0	11.2
Korea	11.1	11.5	6.1	8.7	9.4
Singapore	8.8	11.1	9.2	8.3	9.4
China	10.6	0.8	4.0	S.0	7.6
Hong Kong	1 3.8	7.9	2.3	2.3	6.6
Malaysia	5.2	8.9	8.8	9.4	8.1
Indonesia	3.6	5.7	7.4	7.0	5.9
Philippines	4.4	6.3	5.6	2.5	4.8

Source: NESDB for Thailand, Asian Development Bank for other countries.

Table 2. Average annual growth of real GDP and real per Capita GNP in Thailand, 1960-96

Year	Real GDP	Agriculture	Industry	Services	Real per Capita GNP
60-65	7.2	4.8	11.5	7.2	-
65-70	8.6	6.0	10.4	9.5	-
70-75	5.6	3.8	7.3	5.6	2.9
75-80	7.9	4.0	10.6	8.2	5.3
80-85	5.6	4.9	5.0	6.3	3.5
86	4.5	0.2	7.1	4.6	2.6
87	9.5	0.2	12.8	11.1	7.7
88	13.2	10.2	17.4	11.6	11.4
89	12.0	6.6	16.2	11.1	10.5
90	10.0	1.8	15.8	10.0	8.5
96	N.A.	3.6	9.3	7.9	N.A.

Note: Figures from 1970 are based on the New Series of National Accounts.
Source: NESDB.

By 1996, GDP at current prices, estimated at US$ 186 billion was double that in 1990. The success story of economic growth in Thailand during the past three decades was mainly the result of rapid industrialization, export orientation, and production diversification. The economy has undergone dramatic structural change due to the rapid economic growth of the period. Once an inward-looking, import-substitution economy relying on exports of agricultural products and raw materials, Thailand has been so transformed that the share of GDP from agriculture has declined from 26% in 1970 to 12% in 1991. At the same time, the share of GDP from industry has risen from 16% in 1970 to 26% in 1991, and is expected to continue rising throughout the decade. The magnitude of the changes is even more impressive if one looks at the shifting structure of exports. In 1990, manufactured goods accounted for 75% of total merchandise exports, a dramatic increase compared with the tiny 6% share in 1970, while the export role of agriculture decreased from 70% in 1970 to 23% in 1990 (Bank of Thailand, 1991). Prior to 1996, the tourism sector of Thailand had also experienced rapid growth for a decade. International tourism in 1983 earned US$ 1,002 million, or 17% of the total value of exports, and US$ 4,640 million in 1990, a fourfold increase within only seven years. Total earnings from tourism including domestic tourism are clearly much higher.

There is little doubt that the dramatic economic growth in Thailand during the period from 1960 to 1996 brought about significant benefits to the Thai population in terms of job opportunities, higher real income, more consumption goods, and ultimately better living standards. Yet, amidst the affluence generated by industrialization, complex environmental problems have developed. Signs of serious stresses have been evident in many natural resource systems and environmental media. The linkage between economic development and the adverse changes in the state of the environment have been a most important concern of both the public and the government, and both national and international development organizations. Although policies and legal measures exist for environmental protection and conservation, they are often inefficient or ineffective. To sustain a development process, information and analyses on the state of the environment, the linkages between adverse changes in the state of the environment and economic development, and the role of government and non-governmental organizations as well as environmental education are of utmost importance for planners and decision makers.

In this paper, a systematic and holistic framework of analysis is used to examine the linkages between major environmental issues and economic development in Thailand. The paper also evaluates performance and policy outcome of the Thai government and the role of environmental education and non-governmental organizations in dealing with the environmental issues so as to promote sustainable development. The second section of the paper reviews the current state of the environment in Thailand and the linkages between economic development and environmental problems to provide a background for the subsequent discussions. In the third section, I examine the role of the Thai government in environmental management and sustainable development. The fourth section analyzes the role of environmental education and NGOs in Thailand. Finally, the fifth section of the paper offers concluding remarks regarding the formulation and implementation of environmental policy and environmental education for sustainable development in Thailand.

Economic Development and Environmental Problems

In this section, the state of the environment and changes in its quantity and quality due to rapid economic growth in Thailand during the past three decades are reviewed. The linkages between

Table 3. Changes in forest land area of Thailand, 1960-1993

Year	Forest Area (rai)	As % of Country Area
1960	173,176,250	54.00 %
1973	138,925,625	43.32
1976	124,010,625	38.67
1978	109,515,000	34.15
1983	94,219,349	30.01
1985	94,285,000	29.40
1988	89,877,182	28.03
1991	85,436,284	26.64
1993	83,450,625	26.02

rai=1600 m^2
Source: Royal Forest Department, 1995.

economic development and environmental problems are also analyzed.

Forest resources

Thailand, like other developing countries in the tropics, has been facing a serious deforestation problem. The country's natural forest areas have been cleared at an average rate of 3 million rai per year during the past three decades. As a result, the forest areas decreased more than half between 1960 and 1993, from 173.2 to 83.5 million rai (Table 3). Most of the cleared land was turned into agricultural land but about 30% was left fallow. Despite various efforts by the government to solve the problem, deforestation continues. In 1989 logging was officially banned, but land clearance for agriculture and illegal logging still continued at an average rate of 2 million rai per year during 1991-1993 (Table 4). At this rate of forest clearing, it is estimated that Thailand will lose all forest land within 29 years.

Currently, forest lands account for about one fourth of the country's total land area. Half of this forest land is in the Northern Region and only a small portion in the Central Region (Royal Forest Department, 1995). Most of the remaining forest lands are in steep mountainous or watershed areas. Destruction of watersheds and loss of biodiversity are, therefore, increasingly recognized as more important than the mere shortage of timber and fuel wood. In addition, it was found that an expansion of agricultural land into

Table 4: Rate of deforestation in Thailand in various periods, 1973-1993

Period	Area Loss (Mrai)	Deforestation Rate (Mrai/year)
1973-76	14.6	4.8
1977-78	14.5	7.3
1979-82	11.6	2.9
1983-85	4.7	1.5
1986-88	3.4	1.1
1989-91	4.9	1.7
1991-93	2.0	1.0

Mrai=$(1600m^2) \times 10^6$
Source: Royal Forest Department, 1995.

"marginal" areas where yields are lower raises the average cost of food crop production for the whole country (Panayotou and Parasuk, 1990). In order to protect these pristine natural forest areas and biodiversity, the Thai government has initiated a number of conservation policies. One major policy involves an expansion of protected forest areas of mainly national parks and wild life sanctuaries. As a result, the area of national parks increased 40% during 1987-1994. At present, there are about 55 million rai of protected forest areas which account for 69% and 15% of the total forest and total land areas of the country respectively (Supachit, 1994a). However, the government budget allocated to protection of national parks and other protected areas has not increased proportionately with the increase in area. The expansion of protected areas has thus been accomplished at the expense of the effectiveness of protection. Encroachment, wildlife poaching, and illegal logging in all types of protected forest areas in Thailand are common.

Biological diversity

Thailand's biological diversity is a world as well as a national heritage. The four main types of habitat with the greatest degree of biodiversity are tropical forest ecosystems, mangrove forest ecosystems, fresh water ecosystems, and coral reef ecosystems. The tropical terrestrial and aquatic ecosystems are rich in flora and fauna. More than 10% of the world's known animal species or 4,253 out of the total 41,600 species are found in Thailand (Table 5).

Table 5 Number of animal species Found in Thailand and the World

Type	Thailand	World
Freshwater Fish	650	20,000*
Saltwater Fish	2,000	
Amphibians	107	2,500
Reptiles	298	6,000
Birds	916	8,600
Mammals	282	4,500
Total	4,253	41,600

Note: * Fresh and saltwater combined.
Source: Science Society of Thailand.

In recent years, the biodiversity of most ecosystems in Thailand has been declining in terms of both species number and population size due to a number of causes. Over-harvesting, illegal poaching and trading of wild species of animals and plants, natural habitat destruction, outdated laws, and lack of resources for enforcement are major causes. About 143 animal species found in Thailand are under threat of extinction (Table 6). About 111 species of flora and fauna are listed in the endangered category (Table 7). Over 60% of Thailand's coral reefs are in poor or fair condition. Coral reefs in some areas could be permanently lost. The deterioration is caused by increasing coastal sedimentation and pollution, reef blasting, destructive fishing practices, anchoring on reefs, and over-harvesting of fish and shellfish.

Land and soil

Thailand has relatively abundant cultivable land which represents about 65% of a total land area of 320.7 million rai. The average agricultural land holding is about 27 rai. Only about one fourth of the agricultural land is irrigated and is mostly under paddy fields (Table 8).

Another quarter is devoted to field and tree crops. The rest of the land is unsuitable for agricultural purposes unless investment in conservation measures is incorporated. Soil erosion, salinity, and acidity affect about half of the country's agricultural land

Table 6. Endemic species in Thailand

Type	Number
Freshwater Fish	70
Saltwater Fish	50
Amphibians	13
Reptiles	31
Mammals	8
Total	143

Source: Science Society of Thailand.

Table 7. Endangered species in Thailand

Type	Number
Fish	12
Amphibians and Reptiles	12
Birds	41
Mammals	40
Insects	3
Plants	3
Total	111

Source: Science and Technology Research Institute of Thailand.

(Department of Land Development, 1993). Land misuse and conflict in land use were less in the past when population pressure was less intense. At present, however, as much as 30 million rai of arable land is misused. The Department of Land Development reported in 1990 that moderate and severe soil erosion took place on about 108 million rai of agricultural land. Under-utilization as well as unnecessary fallow land are also evident. The fertile arable land has been converted into residential, commercial, and industrial centers. The trend can be seen in such large cities as Bangkok and Chiang Mai. These settlements as well as industrial centers have become the

Table 8. Percentage of agricultural land area by category of land capability

Land Capability	Region					
	North	North-east	Central	East	South	Total
Field Crop/Tree	6.11	9.57	3.80	1.95	4.60	26.03
Paddy Field	5.12	12.67	4.64	1.74	2.03	26.20
Unsuitable for Agriculture	4.59	6.60	1.05	1.74	1.67	15.65
Forest Conserved	17.05	3.76	4.00	1.25	5.28	31.34
Wet Land	0.19	0.31	0.06	0.02	0.20	0.78
Total	33.06	32.91	13.55	6.70	13.78	100.00

Source: Department of Land Development.

sources of various waste products, many of which cause deterioration in land and soil quality.

Insecurity of land ownership is another fundamental problem affecting the utilization of land resources. More than 40% of agricultural land is occupied and farmed without legal land titles. This factor deprives farmers of both access to credit and incentive to improve and properly manage their land. Moreover, the issues of landless agriculturists, rural poverty, and forest encroachment are intertwined. As much as 80-90 million rai or 38% of the total forest land has been encroached upon by some 10 million landless farmers. Crop yields in newly cleared land are high during the first few years and decline rapidly thereafter. The land thus deteriorates and turns into wasteland. New land is further cleared for cultivation and subsequent soil exhaustion continues (Technical Committee on the Global Environment, 1992).

Water

Competition for water resources, water shortages, and floods have been major problems in Thailand for many years. Although Thailand receives about 800 billion cubic meters of rainfall a year, the nation has made increasing use of its water resources. Water demand is increasing rapidly in the Central Region due to rapid industrialization, urbanization, and expansion of dry-season cropping. The shortage of water for dry season irrigation in the Central Plain and for the piped water supply in Bangkok and other large cities has become critical in recent years. Most of the reservoir sites have already been used. At the same time, many of the remaining water sources are no longer easily drawn on to serve the demands of the Central Plain and Bangkok due to the increasing demand for water in other regions. Ground water has been over-extracted in Bangkok Metropolitan areas for more than a decade. The Metropolitan Water Works Authority (MWWA) currently uses more than 500,000 m^3 of ground water a day for piped water supply. Private industries such as factories, large hotels and housing estates, are also using a large quantity of ground water. This over-extraction has caused serious land subsidence in certain areas of Bangkok and its nearby provinces. Land subsidence rates in areas of Sukhumvit, Phrakhanong, Bang Na, Ladprao, and Hua Mark, have exceeded 10 centimeters a year. Although the Department of Geological Resources has recently implemented higher charges for ground water, the problem will not be solved only by this measure because demand for water has increased rapidly along with the country's high growth rates.

While water shortages occur in the dry season, in the rainy season flooding occurs in many areas of the country including the Bangkok Metropolitan Area. The floods in Bangkok cause significant political and economic impacts and, at the moment, there is no effective prevention measure for the problem. Clearly, Thailand needs to adopt a more efficient and effective approach to water resource management if sustainable development is the country's prime objective.

Coastal resources

Major coastal resources include mangrove forests, coastal and marine fishing and breeding grounds, recreational resources, and use as a waste sink. Like other natural resources, coastal resources in Thailand have been exploited and are in a deteriorating condition due to rapid economic development without regard for environmental concerns and measures.

In 1961 Thailand had about 2.3 million rai of mangrove forest area. In 1993, the mangroves had decreased to about a million rai due to the rapid conversion of mangrove areas for other uses, such as settlement, industries, aquaculture, mining, etc. (Table 9).

Table 9. Changes in mangrove area in Thailand, 1961-1993

Year	Area (rai)
1961	2,299,375
1972	1,954,375
1979	1,795,675
1986	1,227,674
1989	1,128,750
1991	1,112,694
1992	1,096,169
1993	1,054,266

rai=1600m^2
Source: Royal Forest Department, 1995.

Over-fishing is caused by combined factors such as population pressure, modern technology, new economic opportunities, and ineffective legislation and management. Although the total marine fishery production has not yet decreased, there is an increasing percentage of low economic value fish in the composition of marine catches. The share of these low economic value fish increased from 22% in 1971 to 41% in 1988 (Ministry of Agriculture and Cooperatives, 1989/90).

Examples are coconut plantations and shrimp farms in Samut Songkarm and Samut Sakhon, resettlements and industrial sites in Chachoengsao, and fuel wood production along the west bank and Mae Klong River. The area under shrimp farming has increased rapidly during the last decade, from 283, 549 rai in 1986 to 455, 075 rai in 1992 (Table 10). It now accounts for over 90% of aquaculture (Fishery Department, 1990).

Apart from over-fishing and the direct exploitation of mangrove forests, inland discharge is another major cause of coastal and marine resource deterioration. For a long time both industries and tourism development have discharged their wastes into rivers and seas and thus affected many of the coastal resources. In several major tourist areas such as Pattaya, Chon Buri, Kho Samed, Kho Samui, and many other islands, there are inadequate or no sewage systems (see Supachit, 1992). Heavy loads of biochemical oxygen demand (BOD) draining from these areas into the Gulf of Thailand have already reduced the dissolved oxygen (DO) levels in many sections around all

Table 10. Changes in areas of shrimp farms in some provinces of Thailand, 1986-92, in rai

Province	1986	1988	1990	1992
Trat	5,975	8,748	11,382	14,000
Chanthaburi	12,029	38,474	52,898	88,146
Rayong	278	6,326	10,833	9,472
Suratthani	23,098	41,580	55,030	59,540
Nakhonsi-Thammarat	47,220	52,489	62,500	65,019
Songkhla	-	1,169	4,000	18,321
Phatthalung	-	20	120	1,120
Pattani	1,160	2,261	2,800	3,983
Ranong	-	27	800	1,913
Phangnga	-	253	800	3,225
Phuket	145	459	1,100	1,695
Krabi	206	266	380	2,400
Trang	-	356	750	3,372

rai=1600m^2
Source: Department of fishery, 1992.

estuaries to near zero. Heavy metals and pesticide run-off from industrial and agricultural wastes affect the reproductive and the survival rates of fishery resources. Moreover, the harmful impact on fish in their natural habitats can be further passed to aquacultural farms in the same way. Therefore, if these wastes are not properly managed, they can cause permanent loss of valuable marine fishery products.

Ambient quality

The problems of ambient quality are more of a concern for urban areas. Major issues of ambient quality include air pollution, water pollution, solid and hazardous wastes. Industrial, agricultural, and residential areas are significant sources of water pollutants because of an inadequate sewage infrastructure and a shortage of sewage treatment facilities. Untreated waste water from all sources is directly discharged into rivers and canals. Residential waste water discharges are now a serious threat to water quality in Thailand, particularly in urban areas like Bangkok. In 1988 the BOD load from residential sources accounted for 80% of the total BOD load to Chao Phraya River. The Ministry of Science, Technology and Environment reported that while the major rivers in the Northern, Northeastern, and Southern Regions of the country are still in a fair condition, there are signs, such as an increase in the concentration of the total coliform bacteria (TCB) in those water sources, indicating the deterioration of these water sources, particularly in areas of concentrated settlement and industrial development (Table 11).

The untreated waste water discharged from the domestic, commercial and industrial sectors into the surface water sources is the main cause of water pollution problems in these regions. In the Central Region, which is the most populated and industrialized region of the country, the water quality of its major rivers is much poorer than other regions. The water quality of the middle and lower range of the Chao Phraya Rivers is very poor as indicated by the values of DO, BOD, and TCB of the river. The water quality of other major rivers, such as Tha Chin, Mae Klong and Pasak, has also deteriorated rapidly in recent years due to an increase in the use of the rivers as a waste receptacle.

For ambient air quality, the quality of air in Bangkok is usually much poorer than average and considerably worse than in the rural areas. This is due to the concentration of industrial and economic activities, high intensity of energy use, traffic congestion and overcrowding (see Table 12 for trends in the production, import and consumption of energy in Thailand). Ambient air quality has been monitored on a regular basis only in Bangkok, and on a periodic basis in a few other big cities. The average values of all parameters monitored and measured, such as lead, suspended particulate matter, and carbon monoxide have increased significantly since the last decade. In fact, Bangkok has already been placed among the top ten worst cities in the world in terms of air and noise pollution (Khemaphirat et al., 1991). The mean maximum values of carbon monoxide and lead in major Bangkok streets have been high and occasional breaches of the standards do occur. The level of suspended particulate matter in major streets of Bangkok has exceeded the National Ambient Air Quality Standard value (Table 13). Air quality problems also exist in areas where mining and lignite power plants are located, for example, Mae Moh in Lampang Province. However, based on the available information on the changes in level of various air pollutants in areas near major streets in Bangkok, the quality of ambient air is improving (see Tables 14-16). But the 24-hour average noise level in many business areas has increased from 68 dBA to 78 dBA compared with the standard noise level set by the National Environmental Board at 70 dBA (Table 17).

Table 11. Change in quality of surface water sources in Thailand, 1992-93.

Region/River	Standard Value			1992			1993		
	DO	BOD	TCB	DO	BOD	TCB	DO	BOD	TCB
Central Region									
Chao Phraya									
Upper Part	≤6	≥1.5	≥5000	5.6	2.2	210000	5.9	1.7	39700
Middle Part	≤4	≥2	≥2000	3.8	1.7	207777	4.9	2.3	248700
Lower Part	≤2	≥4	-	0.3	8.2	-	1.5	2.7	257700
Thachin									
UpperPart	≤6	≥1.5	≥5000	3.2	3.4	43000	2.9	3.3	84200
MiddlePart	≤4	≥2	≥2000	3.0	5.8	210000	3.6	2.1	109500
Lower Part	≤2	≥4	-	2.2	6.7	765000	2.4	4.2	84300
Maeklong	≤4	≥2	≥2000	4.3	3.0	10000	6.7	1.6	87100
Bang Pakong	≤4	≥2	≥2000	3.8	1.2	250000	4.4	2.3	37400
Pasuk				6.4	1.3	205200			
Sakhaekang							5.3	1.9	133000
Northern Region									
Ping				7.0	1.1	24400	6.4	1.1	67500
Wang				7.6	1.5	20300	6.1	1.1	126500
Yom				6.5	1.6	21700	5.7	1.4	49000
Nan				6.6	1.4	25600	6.3	1.3	83900
Northeast Region									
Chi				7.7	1.8	3000	6.8	1.9	8600
Mun				7.7	3.0	35000	6.6	1.5	7200
Pong							4.4	1.5	7000
Southern Region									
Tapee-Pumduang				6.0	1.4	4500	6.5	1.9	122700

DO=Dissolved Oxygen, BOD=Biological Oxygen Demand,
TCB=Total Coliform Bacteria

Source: Department of Environmental Planning and Policy, 1995.

Table 12. Production, import and consumption of energy in Thailand, 1991-93 (Unit: barrels of crude oil/day)

Category	Quantity			% Change		
	1991	1992	1993	1991	1992	1993
Production	290,212	310,788	331,766	17.00	7.09	6.75
Crude Oil	24,503	26,317	24,694	2.23	7.40	6.17
Condensate	19,211	24,589	26,416	26.92	27.99	7.43
Natural Gas	140,605	150,386	168,747	24.03	6.96	12.21
Lignite	85,894	91,085	95,873	16.33	6.04	5.26
Hydro Power	19,999	8,411	16,036	-8.06	-7.94	-12.90
Imports	372,920	429,249	502,284	5.19	15.10	17.01
Crude & Refined	381,455	435,410	497,875	5.81	14.14	14.35
Condensate (Export)	-17039	-14532	-10227	12.42	-14.71	-29.62
Coal	5,871	6,247	11,704	37.69	6.40	87.35
Electricity	2,633	2,124	2,932	-9.08	-19.33	38.04
Consumption	666,520	721,097	807,049	10.63	8.19	11.92
Crude & Refined	410,794	451,899	509,873	6.42	10.01	12.83
Natural Gas	140,794	150,614	169,011	24.04	6.95	12.21
Coal	5,871	6,247	11,704	37.69	6.40	87.35
Lignite	86,283	91,714	97,438	16.72	6.29	6.24
Hydro/Electricity	22,752	20,623	19,023	-8.09	-9.36	-7.76

Source: Office of the National Committee of Energy Policy

Table 13. Air quality near major streets in Bangkok, 1994 (mg/m^3)

Pollutant	Measured Range	Standards
Carbon Monoxide (8 hr. avg.)	10-22	20
Lead (24 hr. avg.)	0.001-0.006	0.01
Suspended Particulate Matter (24 hr. avg.)	0.2-1.4	0.33.

Source: Office of the National Environment Board.

Table 14. Changes in the concentration of 24-hour-average suspended particulate in areas near major streets in Bangkok, 1992-93 (mg/m^3)

Permanent Monitoring Station	1992	1993	Change
Bang Yeekhan Electricity Sub-station	0.24	0.33	0.09
Phratoonam	0.62	0.73	0.11
Yaowaraj	0.79	0.5	-0.29
Office of National Statistics	0.29	0.23	-0.06
Saphan Khawai	0.28	0.31	0.03
Bangkok Christian Hospital	0.4	0.34	-0.06
Police Department	0.34	0.29	-0.05
Huamark Post Office	0.97	0.83	-0.14
Department of Land Development	0.27	0.24	-0.03

Note: The 24-hour-average ambient standard for suspended particulate matters is 0.33 mg/m^3.
Source: Department of Pollution Control.

Table 15. Change in the concentration of 24-hour-average lead level in areas near major streets in Bangkok, 1992-93 (mg/m^3)

Permanent Monitoring Station	1992	1993	Change
Bang Yeekhan Electricity Sub-station	0.26	0.35	0.09
Phratoonam	0.66	0.68	0.02
Yaowaraj	0.71	0.61	-0.1
Office of National Statistics	0.74	0.37	0.37
SaphanKhawai	0.94	0.35	-0.59
Bangkok Christian Hospital	0.65	0.44	-0.21
Police Department	0.48	0.32	-0.16
Huamark Post Office	1.47	1.11	-0.36

Note: The 24-hour-average ambient standard for Lead is 10 mg/m^3.
Source: Department of Pollution Control

Table 16. Change in the concentration of 1-hour-average carbon monoxide level in areas near major streets in Bangkok. 1992-93 (mg/m^3)

Permanent Monitoring Station	1992	1993	Change
Department of Case Enforcement	2.97	4.98	2.01
Bang Yeekhan Electricity Sub-station	3.63	6.18	2.55
Pratoonam	5.36	4.56	-0.8
Yaowaraj	8.17	5.18	-2.99
Samsean Kindergarten School	8.68	7.43	-1.25
Saphan Khawai	7.83	5.67	-2.16
Bangkok Christian Hospital	8.49	9.04	0.55
Police Department	6.63	3.89	-2.74
Huamark Post office	13.73	9.77	-3.96

Note: The 1-hour-average ambient standard for carbon monoxide is 50 mg/m^3.
Source: Department of Pollution Control.

Table 17. Change in the level of noise in areas near major streets in Bangkok, 1992-5 (decibels)

Point of Measure	1992	1993
Pho Po Ro Building in Chulalongkorn Hospital	77	71
Ministry of Science, Technology and Environment	72	72
Department of Land Transportation	76	78
Odian Circle	68	69

Source: Department of Pollution Control.

Solid waste and hazardous waste

Solid waste has become a more serious problem for the urban environment in the last decade. At present, the Bangkok metropolitan areas alone generate about 6,634 tons of solid waste per day (Table 18). In addition to solid waste, approximately 2 million tons of hazardous waste are produced in Thailand per year (Table 19). Most of this is in the form of heavy metal sludge and solids and these pose significant management and health problems to the society. At present, there is only a small hazardous waste treatment center in the outskirts of Bangkok. The government has approved a plan to develop four more hazardous waste treatment centers in the provinces near Bangkok, but the plan has not been successfully implemented due to NIMBY ("Not-In-My-Backyard") protests by the people in those provinces.

Human resources

The process of rapid economic development affects human resources adversely through the impacts on general and occupational health. Agricultural products are increasingly contaminated with residuals of chemical pesticides (Table 20). Human health in general has been threatened by the worsening conditions of ambient quality. Although the measured levels of many pollutants in the air and water have not exceeded the standards, they could have long term accumulative effects on human health. A high incidence of respiratory health problems is already reported. In 1990, there were about 1 million cases of air-pollution-related respiratory diseases in Bangkok. At the same time, a survey conducted by the Department of

Table 18. Quantity of collected solid waste in Bangkok Metropolitan area, 1980-95 (tons per day)

Year	Quantity
1980	1,966
1982	2,527
1984	2,557
1986	3,738
1988	4,225
1989	4,085
1993	7,000
1994	7,050
1995	6,634

Source: Bangkok Metropolitan Authority.

Table 19. Type and volume of hazardous waste, 1986-91 (tons)

Waste	1986	1991	Annual Change (%)
Oil residues	124,194	219,467	15
Liquid organic waste	187	311	13
Organic sludge	3,737	6,674	16
Inorganic sludge	11,698	19,254	13
Heavy metal sludge	823,869	1,447,590	15
Solvent	19,738	36,163	17
Acid Waste	81,054	125,428	11
Alkaline Waste	21,952	34,235	11
Off-spec products	12	25	22
Aqueous organic waste	116	242	22
Photographic waste	8,820	16,348	17
Municipal waste	7,231	11,787	13
Infectious waste	46,647	76,078	13
Total	1,151,729	1,993,602	15

Source: Engineering Science Inc.

Table 20. Residuals of chemical pesticide in agricultural products

Type	Number of Sample	% of Sample with residual	% of Sample with over-dose residual
Fruits			
Fruit with eatable peel	121	93	13
Fruit with inedible peel	198	62	6
Vegetables	864	34	5
Cereals	105	2	-
Dry Beans	71	30	-
Animal Products	467	74	-

Police found that traffic policemen suffered from health problems such as hearing impairment, high blood pressure, reduced lung function ability, and sinus inflammation (Khemaphirat et al., 1991). A survey conducted by Chulalongkorn University found that 82% of the commuter boat operators in selected canals in Bangkok suffered hearing impairment from the noise generated by their engines, and all operators working longer than 15 years had irreversible hearing loss (TDRI, 1987).

A Holistic View of the Problem

The changes in the state of the environment in Thailand during the past three decades reviewed in the previous subsections indicate that, along with the country's very successful economic growth, Thailand is currently plagued with complex environmental problems. Obviously, the emergence of natural resource depletion and environmental degradation is closely linked to rapid economic growth. The consequences of economic development and environmental change in Thailand are explained below.

First, economic growth has depended upon the exploitation of the natural resource base. The forests, fertile lands, water, and coastal areas have been heavily used for production in the agricultural and industrial sectors either with inadequate conservation measures, or without any at all. This has led to rapid depletion and deterioration of most natural resources through out the country.

Second, industrialization has been concentrated in urban areas, especially the Bangkok Metropolitan Region (BMR-Bangkok and the

Table 21 Expansion of industries in Thailand, 1969-89

Indicator	1969	1979	1989
Industrial GDP	37,578	83,709	195,444
% Share in GDP	24.14	29.29	34.03
No. of Factories	631	19,691	51,500

Note: 1) Industrial GDP is quoted at constant 1972 prices and in million baht; 2) Only factories registered with the Department of Industrial Works are reported.
Sources: Bank of Thailand and Department of Industrial Works.

five surrounding provinces), due to the proximity to markets and ports. It has induced a large increase in the demand for energy and raw materials, and has also generated more wastes and pollution per unit of production than the agricultural sector that it is replacing. During the three decades of rapid industrialization in Thailand, the number of industrial factories has increased from a few hundred to over fifty thousand (Table 21). In 1990, the BMR accounted for over 50% of the 52,000 factories and 23 industrial estates in the country and generated 75% of industrial waste. There are signs that some of the worst industrial polluters are moving out of the inner city district into its satellite provinces The manufacturing sector is by far the largest generator of hazardous waste, accounting for 90% of all such waste in the country (Phantumvanit and Panayotou, 1990).

Third, industrialization has induced urbanization due to the availability of employment in industrial and service sectors. Urbanization further increases energy consumption and pollution. The population of the BMR is currently growing at twice the national average. Migration towards urban areas is likely to continue and will probably accelerate. Major urban environmental problems include the deterioration of the ambient environment and the inability of the public sector to meet the rapidly growing demand for infrastructure by the urban population. Air pollution in Bangkok and other large cities arises from increasing levels of industrial and vehicle emissions. Water pollution arises from emissions of untreated waste water from industries and households. Industries and household refuse are sources of hazardous and toxic wastes in urban areas.

Finally, the expansion of economic activities involving the use of natural resources and public infrastructures is occurring much faster than the policy and legislative development for their control could keep up. This has created serious implications for the productivity of human resources and the whole economy. Furthermore, when most of the natural resource bases in the rural areas are degraded and depleted, poverty and a wider gap of income distribution between the rural and the urban sectors will intensify and complicate the problem.

Most environmental problems basically reflect market failure. As a result government intervention by means of various plans and policies is often seen as necessary. However, governments are bureaucratic by nature, and NGOs as well as community participation have proved to be more efficient and more cost-effective in managing and handling many environmental problems. The roles of NGOs and environmental education as a way to increase the society's environmental awareness and participation are, therefore, important.

In the following two sections, the performance and policies of the Thai government and the roles of environmental education and NGOs in Thailand in gearing the economy toward sustainable development are analyzed.

The Role of the Thai Government in Environmental Management

This section investigates how the Thai government has attempted to deal with the environmental sustainability problems after three decades of rapid economic growth. The analyses focus on the country's national development policies and legal and institutional arrangements.

National development policies

Thailand formulated her first National Economic and Social Development Plan in 1961. The primary objective was economic growth as formally indicated in the plan. At the time the country's rich natural resource base was viewed in terms of its economic potential and was treated as a limitless natural stock contributing to economic development and growth. In addition, the plan adopted a by-sector development approach. Each economic sector, i.e., agriculture, industry, transport, education, etc. was encouraged to do its best. The plan lacked strategies to integrate sectoral development plans for an overall balanced development (NESDB 1991).

Throughout the five year period (1961-1965) of the First Plan, development activities focused on the construction of infrastructure;

particularly roads, railways, and large scale irrigation systems. The purposes were to improve access to rural areas and accelerate the utilization of various natural resources. The government provided various incentives for agricultural land development, commercial logging, and mining. Timber and mining concessions were granted to both foreign and domestic private firms on generous terms.

The succeeding Second and Third Plans (1966-1976) continued to promote growth and economic productivity in the short term. The long term ecological impacts of natural resource consumption received little attention. The general outcome was an expansion of cultivable areas and a rapid decrease of forest lands through out the country. Although the government had a goal to conserve 50% of the total land area of the country as forest land, this was not achieved.

After 15 years of natural resource exploitation for economic growth, depletion of forest resources, deterioration of soil quality, and shortages in water supply became noticeable. The sectoral development approach without integrated management plans for natural resources and the environment had created far reaching impacts on the natural world. The bureaucratic system of the government and the one-policy-for-all development strategy are not efficient in dealing with social and environmental problems. Various environmental problems fall into the same category, but they vary in detail from one locality to another. Each area may require separate treatment to ensure compatibility with local socioeconomic and cultural settings (Supachit 1994c).

The Fourth Plan (1976-1981) contained some resource protection and rehabilitation strategies. They were mostly preparations of groundwork for natural resource planning and management, i.e., problem analyses, compilation of data on natural resource utilization and environmental problems. The government also initiated a number of reforestation programs. However, the government still relied for all policy implementations on its organizations and officials. The bureaucratic nature of the governmental system and organizations prevented the government from carrying out the tasks successfully. Government officials usually cannot remain in an area long enough to identify and understand diverse local resource utilization and environmental problems. The use of a top-down policy and one-policy-for-all strategy to manage a problem which has complex local dimensions caused failure in the implementation of the government's resource protection and rehabilitation plans during the Fourth Plan.

In the Fifth Plan (1981-1986), the government introduced an integrated approach to natural resource development with an aim to increase the efficiency of natural resource utilization and restoration at the local level. The government implemented the idea by introducing a number of projects to improve productivity and the welfare of the people in certain areas. Each project aimed at local socioeconomic development rather than short-term monetary income gain. The Fifth Plan made the government realize that the implementation of a handful of projects could not rescue the situation. Economic development and environmental protection need to be integrated first at the macro level to set a national development goal and a central mission of government operations. Development strategies are crucial if one recognizes that the problems have been increasingly intensified.

The Sixth Plan (1986-1991) marked a turning point in the government's economic development strategies on natural resources and environmental planning. The government realized that the intensified depletion and degradation of the resource base required a total revision of the development perception to one in which the resource base acts as a constraint on economic development. Natural resource conservation and environmental protection were, therefore, explicitly expressed as one major objective of the plan. The Sixth Plan emphasized the development of alternative, non-agricultural sources of income to reduce forest encroachment and dependency of farmers on the increasingly deteriorating soil in forests which had been encroached on. The government also attempted to decentralize natural resource management to the provincial level to promote a sense of ownership, participation, and awareness at the local level. To mitigate environmental impacts from development projects, both private and public organizations are required to submit environmental impact assessments for project consideration and wait for approval to ensure effective impact mitigation.

A change in economic development strategy introduced in the Sixth Plan to mitigate adverse impacts on the natural environmental system has not, however, been matched by effective implementation. At the conclusion of the Fifth Plan, the inability of the government to monitor, control, and regulate its own public sector effectively to follow the new economic development strategy was noted, particularly in relation to natural resources and environmental management. The operations of the public sector have not been in alignment with the principles of sustainable development. The requirement of environmental impact assessments for development projects has been only a rubber stamp for project approval. A number of large scale development projects started construction before the projects were approved. Similarly, many development projects, including those undertaken by the public sector, created far-reaching environmental problems in their operation processes despite their soundly proposed mitigation plans. Project follow-up, monitoring, and enforcement have not been undertaken in practice.

Resource management at sector and project levels by government agencies has also created much stress on natural resource stock and environmental quality. Of note are forest resources, energy, and tourism. Difficulties in defining property rights in relation to resources and the environment have contributed to the emergence of problems in relation to these resources and sectors of the economy. Rapid economic growth has further intensified these problems. The government has faced serious problems in allocating and protecting the country's limited forest lands for environmental quality control. Major obstacles to solving the deforestation problem lie in the management of government agencies as described above. The rapid increase in energy production and consumption has resulted in air pollution in many urban areas (see Supachit, 1994a; Supachit and Pitrachat, 1993b). In the energy sector, a true cost pricing has not been implemented. This resulted in inefficiency in both resource allocation and pollution reduction. Thailand's tourism development policy is evidently costly without environmental protection (Supachit, 1992).

In the Seventh Plan (1991-1996), the government explicitly stated that the country's rapid economic growth during the past decades had exploited much of the country's resource base and in the absence of effective resource management and environmental protection efforts, the depletion and degradation of the resource base had now intensified to an alarming level. The government realized the necessity to revise its former development perception to one in which the resource base is viewed as limited and might constrain economic development. More efforts had to be made to monitor the environmental effects of socioeconomic policies, and a systematic framework for environmental policy had to be formulated. The formulation of the Seventh Plan involved broad-based participation. It was based on the cooperative effort of all sectors, including government agencies, state enterprises and universities, the private sector, and non-governmental organizations. As a result, the policies in the Seventh Plan reflected a liberalization process in many respects.

The Seventh Plan had three principal objectives: first, to sustain the country's economic growth at an appropriate level, with stability;

second, to promote more equitable income distribution and rural development; and third, to improve the quality of human resources, life, natural resources, and the environment. Policies on natural resource utilization and environmental protection differed from those in the previous plans on a number of counts. Fundamentally, the Plan focused on five major areas of environmental management policy: natural resource management, environmental quality, energy and environment, industry and environment, and urbanization and environment. The Seventh Plan set definite targets for these areas to ensure the effectiveness of improving the quality of natural resources and the environment throughout the country. Examples included targets for the rehabilitation of water quality of the lower Chao Phraya and Tha Chin Rivers, management of industrial hazardous waste, control of air pollution along major streets in Bangkok, protection of forests in watershed areas, reforestation, conservation of coastal areas, and the protection of tropical coral reefs. In addition, the plan recognized that these tasks are beyond the scope of the government alone.

All programs have various forms and methods of drawing contributions from all sectors of the economy. In environmental quality management, pollution control and waste treatment projects must be paid for by polluters and generators of waste based on the "polluter-pays principle" rather than be subsidized through the use of national tax revenues. The amount of government budget allocated to environmental protection activities and plans of various ministries has also increased (Tables 22, 23).

At the same time investments in environmental facilities have been initiated. At present, there are at least seven central sewage treatment plants under construction in Bangkok areas (Table 24). In the area of natural resource management, local people will be called to take part in the management of terrestrial and mangrove forests, biological diversity, land, and water. Non-governmental organizations will be encouraged to assist the government in this endeavor by mobilizing rural people to participate in natural resource management programs.

The Eighth Plan (1997-2003) addressed explicitly the sustainable development objective and human resource development has been seen as the most important factor to achieve it. Decentralization of government administration and tax-revenue collection, privatization of government and semi-government organizations, and rights of the public to information are planned. Various arrangements to reform existing institutions and establish new ones have also been made. All institutions and classes of people in Thailand were, for the first time,

Table 22. Government budget allocated to the departments in charge of environmental policy of the MOSTE (millions of Baht)

Department	1992	1993	1994	1995	1996
Policy and Planning	85.1	1078.7	1226.5	15691.5	19160.8
Pollution Control	91.9	200.1	408.9	844.9	101.6
Environmental Quality Promotion	80.1	116.3	171.7	52.1	153.8

Source: Ministry of Science Technology and Environment (MOSTE).

Table 23. Government budget allocated to environmental protection by Ministry (Allocated Annual Budget, millions of Baht)

Ministry	1992	1993	1994	1995	1996
Agriculture	5621.6	7164.3	9052.2	16020.9	2905.7
Public Health	71.4	242.5	352.2	229.5	46.0
Industry	395.5	274.2	562.7	593.9	941.1
University Affairs	11.9	2.8	21.9	43.7	0
Science	5253.6	5587.8	6436.9	29139.0	31694.1
Interior	3132.9	6151.9	8703.4	32273.4	7074.3
Education	115.1	143.6	177.4	333.4	138.4
Total	14602.0	19567.1	25306.7	78633.8	42799.6

Source: Ministry of Science Technology and Environment.

Table 24. Central sewage treatment plants in Bangkok Metropolitan areas

Project	Year Completed	Area Covered (m²)	Capacity (m³/d)	District Covered
See Phraya	1996	2.6	30,000	Samphanthawong
Ratanakosin	1995	4.0	25,000	Phranakorn
Central System I	1996	37.0	350,000	Pomprab, Samphanthawong
Lumpini	N.A.	N.A.	180,000	Prathumthani, Rachathewi
BMA	N.A.	N.A.	170,000	Phranakorn, Dusit, Phrayathai, Huikwaung
Yannawa	N.A.	28.5	200,000	Yannawa, Sathorn, Bangrak, Bang Kholeam
Nongkham-Phasee Chareon	N.A.	40.0	157,000	Nongkham, Phasee
Ratburana	N.A.	41.0	7,000	Ratburana

Note: BMA stands for Bangkok Metropolitan Authority.
Source: Division of Policy and Plan Bangkok Metropolitan Authority.

allowed and encouraged to contribute their ideas of development strategies to the drafting of the Eighth Plan. The current national development plan, therefore, marks another turning point in the role of the Thai government.

The above review of Thailand's development policy and efforts to gear the economy toward ecologically sustainable development indicates that over the first three national development plans, the Thai government has been very successful in guiding the economy toward economic growth and modernization. Natural resources were treated as unlimited in terms of quantity and quality. A sectoral development approach was employed without any integrated management plans for the system of natural resources and environment. Economic growth thus created far reaching impacts on the natural system. Political centralization, the bureaucratic system of the government, and the one-policy-for-all development strategy are not efficient in dealing with complex social and ecological problems. Taken together, they apparently intensified the problems. In the later development plans, although the government has recognized the problems and attempted to improve the plans by adopting several solutions to the problems of natural resource depletion and environmental deterioration, most solutions were ad hoc, segmented, and insufficient.

During the last decade, the government has undergone several philosophical shifts on natural resource planning and policy. Although these shifts have not been matched by implementation, they marked a new direction of development strategy. The Seventh Plan provided more responses to the ecological sustainability problems. Some of the changes in policy and management of natural resources and environmental quality appeared promising in terms of their effectiveness. Of note are mandates to promote public awareness and participation in pollution abatement and natural resource conservation. The principle of "polluter pays" has been used as a mean to enforce environmental standards. The plan, however, lacks a number of effective pollution control mechanisms. Mitigation of environmental impacts from economic development activities is still based on a project-level approach. Industrial hazardous waste control still focuses on end-of-pipeline control rather than waste minimization in the production process. In addition, the "polluter pays" principle has been adopted without a clear enforcement strategy. In the present Eighth Plan, the government has determined to play a larger role in implementing more effective policy measures to guide the economy towards sustainable development in both the public and private sectors. The idea of sustainable development issues has been integrated into human resource development, the decision-making structures of the government, and macroeconomic policy.

Legal and institutional arrangements

The political system in Thailand is still centralized. The governor of each province is appointed by the central government in Bangkok. Decision making on development policies, implementation strategies, as well as institutional arrangements all come from the cabinet and the bureaucracy. Government agencies are responsible for policy implementation, which includes information collection, data compilation, monitoring, control, and enforcement. Although the concepts of decentralization and privatization have been accepted by a number of government organizations, the process of change has been slow in almost every organization.

Almost all programs of natural resource rehabilitation and environmental protection fall into the hands of the government. The shortcomings of the market in promoting economically optimal (efficient) or socially desirable (equitable) outcomes when technological problems are present have justified public policy intervention in many countries, including Thailand. The management of land and forest resources, water resources, as well as various aspects of environmental quality in Thailand are mostly handled by government agencies. Land and forest resources are typical examples in this regard.

There are currently 11 laws that directly or indirectly govern the management of land resources in Thailand. They are: 1) the 1941 Forest Act, 2) the 1954 Land Code, 3) the 1960 Wild Animal Reservation and Protection Act, 4) the 1961 National Park Act, 5) the 1964 National Forest Reserves Act, 6) the 1968 Land Settlement Act, 7) the 1974 Land Consolidation Act, 8) the 1992 Improvement and Conservation of National Environmental Quality Act, 9) the 1975 Town and Country Planning Act, 10) the 1975 Agricultural Land Reform Act, and 11) the 1983 Land Development Act. The various government agencies dealing with land administration operate under different ministries. The government land agencies have been operating without adequate coordination due to the lack of a national land policy. This creates difficulties in law enforcement which may arise from the inconsistencies in the acts and the conflict between the governmental agencies responsible (Supachit, 1994a). In addition, most of them are facing similar problems in administration, i.e., lack of funds, manpower, and government support. There is a need for the reorganization and restructuring of the land administration system in order to improve administrative efficiency and coordination among the responsible agencies.

The Seventh, as well as the Eighth Plans have introduced a major change in legal and institutional arrangements governing environmental protection. A new comprehensive environmental law was recently enacted. It empowers policy and planning agencies with the right of enforcement, and it decentralizes control over the environment to provincial and local government. The law recognizes the public's right to know and to participate in environmental affairs, as well as the role of the private sector and nongovernmental organizations in environmental rehabilitation. The law has also established three organizations at the departmental level in the Ministry of Science, Technology and the Environment which are responsible for the nation's natural resource and environmental management policies. Apart from the national environmental law, three other major laws on related issues were also enacted. These include laws on public health, industrial firms, and hazardous waste. These new laws are important for ecologically sustainable development although more time is needed to develop enforcement strategies.

Thailand's recent changes indicate the government's commitment to more sustainable development policies in the future. Nevertheless, the government still has a number of important things to do before it can efficiently and effectively address environmental and development concerns simultaneously in planning and policy making. Three suggestions are offered below.

First, the country must have an account of information on changes in quality, quantity, and value of natural resources and the environment over time. The concept and framework of environmental and natural resource accounting (ENRA) can play a major role in this regard. Interest in using this approach to correct real national income or reflect environmental costs and concerns exists among responsible officials although serious attempts and plans have not been initiated. They recognize that environmental and natural resource accounting provides the means by which resource depletion and degradation may be tracked and evaluated, so that the true cost of development programs can be better estimated. At the moment, direct ENRA research projects have not been initiated; there are only ENRA-related studies which are relatively small and mainly academic and theoretical in nature (Supachit and Pitrachat, 1993; Supachit and Limprayoon, 1993).

A feasibility study of integrated environmental and economic accounting was initiated in 1992 as part of the broader efforts of the World Bank to assess the sustainability of economic growth and development in Thailand (Bartelmus and Tardos 1992). The objectives of the mission were: 1) to assess the feasibility of integrated environmental accounting; 2) to compile tentative estimates of the system of integrated and environmental accounts (SEEA) for the period 1970-90; and 3) to make suggestions on how to establish a country project on the implementation of the SEEA. The study took only a total of five weeks, but the mission was able to accomplish a considerable amount of work. It identified potential sources of data and information needed to implement an integrated environmental and economic accounting for Thailand and made rough estimates of some of the necessary parameters in order to assess the feasibility of the future plan. The report pointed to the problem of insufficient data and discussed some major drawbacks of the data set obtained. Nevertheless, the research team anticipated the feasibility of conducting a country project for Thailand and suggested some ideas for carrying it out.

Second, the principle of economic efficiency has not been adequately taken into account in the design and adoption of policies and measurements for natural resource rehabilitation and environmental improvements. Most decision making in this area is still dominated by scientific and technological considerations alone. A recent study of pricing of non-leaded gasoline shows that the government should be able to control air pollution in Bangkok with more efficient alternatives (Supachit and Pitrachat, 1993). In some instances, for example, in the case of water pollution control in the Lower Chao Phraya River and air pollution control for the power plants in Mae Moh, Lampang Province, the government has not attempted to identify a least-cost alternative because resources are limited. The government will have to pay more attention to choosing the most efficient or least-cost alternative as a remedy so that the portion of resources saved can be allocated to other necessary development activities and policies.

Third, although one principal justification for public policy intervention is market failure, this rationale is only a necessary, not a sufficient, condition for policy formulation or for government intervention. It does not follow that whenever laissez faire falls short government, interference is expedient. The inevitable drawbacks of the government may, in any particular case, be worse than the shortcomings of private enterprise. Policy formulation requires that the shortcomings of market outcomes be compared with the potential shortcomings of non-market efforts to provide remedies. In addition, other institutional arrangements may be more capable of coping with the problems. In general, the supply of non-market activities is characterized by several distinctive attributes that may also lead to non-market failures. Non-market outputs are often difficult to define and measure as to quantity or to evaluate as quality. An absence of sustained competition to government authority further contributes to the difficulty of evaluating the quality of non-market outputs. In addition, there is no effective bottom line and termination mechanism for the performance of the government agencies. In the areas of natural resource management and environmental protection, a large number of parties and individuals involved in the problems indicate a considerable transaction cost of intervention. Experience in Thailand and elsewhere shows that, in many cases, the private sector, non-government organizations, and local communities can manage the problem more cost-effectively (see Supachit, 1994c). This fact well explains the recent movement toward people participation and the use of economic incentive approaches in natural resource management and pollution control respectively. Choosing a least transaction cost implementation strategy is, therefore, crucial to the success of the intervention.

Environmental Education and NGOs in Thailand
The Role of Environmental Education

The emergence of environmental problems in Thailand in the 1970s led to the introduction of environmental education in formal education in 1978. It started at all levels: elementary, secondary, high school, vocational, teachers' college, university under-graduate and graduate levels. The objectives of providing environmental education are: 1) to develop understanding and awareness of, and responsibility for the environment among the students, 2) to provide a concept of living in harmony with the environment and with quality of life, and 3) to encourage student participation in environmental protection and restoration.

At elementary, intermediate, and high school levels, the concept and contents of environmental education have been integrated into various subject groups. For example, a unit entitled "Environment" that teaches about plants, animals, water, air, the earth, the sun and stars was incorporated into the "Experience Build-up" and "Habit Build-up" subject groups for the elementary level. A similar approach is used at the intermediate level but the content covers more details and introduces the pupils to more complicated facts about the world and various major natural resources. To increase the pupils' learning experience, school and community activities such as environmental camping, bird watching, clean-up days, forest conservation, and planting trees and flowers, have also been arranged on various occasions throughout the school years. For vocational, teacher college, and university under-graduate levels, an individual subject approach is used. Examples are Natural Resources and the Environment, Energy and the Environment, Safety and Industrial Pollution Control, Man and Environment, Life and the Environment, Environmental Science for Teachers, Ecology, Man and Society, Man and the Physical Science, Man and Biological Science, and so on.

Most universities offer masters' level degree courses in Environmental Sciences, Environmental Social Science, and Environmental Education. These are offered by conventional as well as newly established environment related faculties and departments. After the first environmental related masters' program, "Technology of Environmental Management" was offered at Mahidol University in 1975, many other masters' courses on the environment were developed and offered. These include masters' programs in Environmental Education, Environmental Social Science, Environmental Health, Environmental Chemistry, Environmental Biology, Environmental Engineering, Environmental Sciences, Environmental Management, Natural Resource Management, Appropriate Technology, Information Systems for Environmental Management, Environmental Impact Assessment, Waste Water Engineering Science, Human Settlement, and Urban Planning. The most recent ones include Environmental Economics and Environmental Law. Some universities, such as Thammasat University and Mahidol University offer Environmental Sciences programs at the bachelors' level as well.

In the non-formal education sector, environmental education has been offered in various forms. The Department of Non-formal Education of the Ministry of Education offers a few subjects on the environment at various levels. Governmental organizations such as the Office of Environmental Policy and Planning (OEPP), Department of Pollution Control, Department of Environmental Quality Promotion, and Department of Religion play a major role in collecting data on the environment using modern information and communication technology. They disseminate these data and promote good public relations and information about environmental problems and protection. They also provide training courses, workshops, seminars, and research grants on monitoring, inventory, and other environmental issues aim at increasing the awareness and responsibility of the general public on the environment and promoting participation in environmental quality protection. The government also launched the "Protect the Environment Campaign" and other related information dissemination on environmental protection nationwide in 1990.

In brief, environmental education, both formal and informal, has been introduced in Thailand for almost two decades. The continuing deterioration of environmental quality and the environmental education provided through the formal and non-formal sectors have significantly contributed to an increase in environmental awareness of the Thai people. The first half of the 1990s marked the peak in environmental movements and campaigns in Thailand. According to Prapat Pintopteng (1994), the number of protest demonstrations across the country reached 739 in 1993 and 754 in 1994. An environmental survey conducted in 1994 shows that Thai people in both rural and urban areas have a strong interest in environmental problems and are aware of the importance of environmental protection. Natural/environmental deterioration was ranked as the most serious problem. This survey also shows that as many as 60% of the respondents in Bangkok Metropolitan Areas gave precedence to "environmental protection" over "economic development". The majority of the respondents felt that their lives and health are significantly affected by the adverse environmental quality.

Environmental education at tertiary levels which started in the 1970s, has contributed to Thai society more in terms of supplying man-power and expertise. Although still inadequate, Thailand possesses a number of environmental scientists, engineers, educators, managers, advocates, and economists. They are currently working in various governmental, non-governmental, and private organizations.

It is expected that public awareness of the issue of the environment and economic development will continue to increase with the influence of environmental education. Nevertheless, if the Thai economy wants to achieve its sustainable development objective, public awareness of environmental protection in Thailand must be transformed into effective public participation. Currently, public participation in environmental management is still limited due to various factors, and environmental education is important in dealing with these. Environmental education in Thailand has contributed significantly to an increase in public environmental awareness. Nevertheless, the Thai economy has reached the point where environmental awareness must be translated into more active public participation in environmental management. Environmental education in Thailand as well as other socio-economic and political institutions should be improved to bring this about. It must penetrate to the level of the family and concentrate on hands-on experience and practice to a greater extent than before.

The role of NGOs

At present, there are more than 170 NGOs in Thailand. As many as 134 organizations address social issues induced by main stream economic development strategy and are engaged in political movements and development activities related to problems of prostitutes, aids, drugs, child labor, child prostitution, homeless children, rural development, poverty, etc.. Recently, most NGOs have also extended their interest and activities into problems of natural resource management and pollution control due to the widespread nature of the problems and their close links with social and economic development.

Most NGOs consist of young educated middle class people who are aware of the society's various socio-economic and political problems and are eager to take part in solving them. Many have gone to the rural areas and worked closely with the communities. NGOs in Thailand, like those elsewhere, are watch-dog organizations. They also lead or support social campaigns and movements. Often, university academics join and guide the NGOs.

Major NGOs in Thailand includes organizations such as the Green World Foundation, Thailand Wildlife and Vegetation Protection Foundation, Komon-Keemthong Foundation, Think Earth Foundation, Magic-Eye Society, Suub Nakhasathiean Foundation, Duang Pratheep Foundation, and Forum of the Poor. NGOs in Thailand have organized seminars, workshops, annual meetings, campaigns, movements, and music festivals for the public. They have played an important role in creating or increasing public environmental awareness. Unlike governmental organizations, NGOs can work at the grassroots level in rural communities. They also have more time to identify and understand the problems of these communities. Unfortunately, NGOs lack the necessary resources for alleviating the problems. They can only advocate alternative development strategies and stimulate campaigns and movements. One major obstacle to NGOs carrying out their job is their lack of access to data and information from most public authorities and organizations.

In the past, the relationship between the NGOs and government organizations was more concerned with confrontation than cooperation. It has become more positive recently. In the Seventh and Eighth Plans the government has promoted the participation of the NGOs in environmental protection and management. NGOs have been invited to observe or be represented on various committees which are responsible for environmental issues and programs. A budget has been allocated to support the NGOs' activities. However, the government still has to improve the criteria for giving financial support to the NGOs. At present, it only provides financial support to NGOs which register with the Department of Environmental Quality Promotion. Many NGOs do not register because they cannot afford the high registration fee and consequently they are not eligible for government support.

As elsewhere, financial support is the key to the future of NGOs in Thailand. In the past, most NGOs in Thailand received financial support from foreign organizations. This source of funds has become more limited since Thailand reached a higher stage of development.

The survival and performance of NGOs in Thailand depends much on how well they are able to get funds from other sources or by other means. A few NGOs have conducted fund raising campaigns, although not very successfully. Some have started to produce green agricultural products or handicrafts in partnership with the farmers and other rural people.

Concluding Remark

This paper has attempted to show that economic development and the quality of the environment are closely related to each other. Economic development without concern for the environment and without proper planning and investment in environmental protection will certainly result in an unsustainable social and economic system. Although in the past the Thai government was not able to respond to the problem effectively due to a number of factors analyzed in this paper, the situation has improved. Judging from its latest national development plan, it is evident that the role of the Thai government in recent years has become one which increasingly devotes its attention to reducing environmental degradation, preventing further environmental damage, and promoting conservation of natural resources and environmental sustainability. The main problems, however, lie with the lack of political will to implement such environmental policies, the conflicts of interest among policy-makers and various sectors of the economy, and the difficulties in enforcing many environment laws, both old and new.

It is in this regard that environmental education and the NGOs have played an important role in increasing public awareness of environmental problems and environmental quality protection. The linkage between economic development and the adverse changes in the state of the environment has been a most important concern of the public and NGOs. It is expected that environmental improvement in Thailand will become even greater in the future when public awareness of the issues further increases and more public policies are subjected to public scrutiny and evaluation. Furthermore, if the Thai economy wants to achieve its objective of sustainable development, it is necessary that public awareness of environmental protection in Thailand be transformed into effective public participation in environmental management through various institutional reforms and the promotion of a more democratic system.

References

Arbhabhirama, A., et al., 1988. *Thailand Natural Resource Profile*. Oxford University Press, Singapore.

Asian Development Bank. 1991. *Asian Development Outlook 1991*. Asian Development Bank, Manila, Philippines.

Bartelmus, P. and A. Tardos. 1992. Integrated Environmental and Economic accounting for Thailand: A Feasibility Study. A draft report submitted to the Department of Economic and Social Development, United Nations.

Khemaphirat, B, et al. 1991. Air and Noise Pollution: The Answers are with the People. In: *Environment '91. Papers for the Second Annual Seminar on Natural Resource and Environmental Conservation*, June 15-16, Bangkok, Thailand.

Phantumvanit, D and T. Panayotou. 1990. *Industrialization and Environmental Quality: Paying the Price*. Thailand Development Research Institute Foundation, Bangkok.

Fishery Department. 1990. *Marine Fishery Census*. Fishery Department, Ministry of Agriculture and Cooperatives, Bangkok.

National Economic and Social Development Board NESDB, Office of the Prime Minister. 1991. *Summary of the Seventh National Economic and Social Development Plan 1991-1996*. NESDB, Bangkok.

National Economic Development Board NESDB, Office of the Prime Minister. 1966. *National Income of Thailand, 1966*. NESDB, Bangkok.

Nishira, Sigeki et al. Ed. 1997. *Environmental Awareness in Developing Countries: the case of China and Thailand*. Institute of Developing Economies, Tokyo.

Office of Policy and Plan, Bangkok Metropolitan Authority. 1995. *A Development Plan for Bangkok Metropolitan Bangkok*. Bangkok Metropolitan Authority (in Thai).

Office of Technical Policy and Planning, Ministry of Science, Technology and Environment. 1995. *A Report on Situation of the Quality of the Environment*. Bangkok: Ministry of Science, Technology and Environment.

Office of Technical Policy and Planning, Ministry of Science, Technology and Environment. 1995. *Report on Environmental Quality B.E. 1992-1993*. Office of Technical Planning and Policy, Ministry of Science, Technology and Environment, Bangkok (in Thai).

Panayotou, T. and C. Parasuk. 1990. *Land and Forest: Projecting Demand and Managing Encroachment.* Thailand Development Research Institute Foundation, Bangkok.

Prapat Pintopteng. 1994. A survey report appeared in Sayam Rat Subpradawijarn, Chabap 26, May 27, 1994, p. 20.

Royal Forest Department. 1995. Changes in Forest Land Area of Thailand during 1960-1993. In *Report on Environmental Quality,* B.E. 2535-2536. Office of Technical Planning and Policy, Ministry of Science, Technology and Environment, Bangkok (in Thai).

Supachit Manopimoke. 1992. *The Environment in the Tourist Economy: A Case Study of Pattaya.* Thailand Development Research Institute Foundation, Bangkok.

Supachit Manopimoke. 1994a. Thailand's Natural Resources and Environment in the Year 2000. In *Proceedings of the Seventeenth Annual Symposium, 23-24 March 1994,* Faculty of Economics, Thammasat University, Bangkok (in Thai).

Supachit Manopimoke. 1994b. Environment and Natural Resource Accounting: the Thai Experience. Paper presented at the International, Workshop on the Contributions to Policy of Environmental and Natural Resource Accounting, the Philippines, 16-21 January, 1994.

Supachit Manopimoke. 1994c. Natural Resource Management: An Issue of Local Dimension. In *Proceedings of the International Experts Meeting for the Operationization of the Economics of Sustainable Development,* the Philippines, 28-30 July, 1994.

Supachit Manopimoke and N. Limprayoon. 1993. Air Pollution Control: A Policy Analysis for the Unleaded Gasoline Pricing in Thailand. In *Proceedings of the Third Mahidol-Thammasat Annual Symposium, 19-21 May, 1993,* Mahidol University, Bangkok (in Thai).

Supachit Manopimoke. and K. Pitrachat. 1993. Incorporating Natural Resource Degradation Cost into Economic Growth Indicator: a Framework for the Thai Agricultural Sector. In *Proceedings of the Third Mahidol-Thammasat Annual Symposium, 19-21 May, 1993.* Mahidol University, Bangkok (in Thai).

Technical Committee on the Global Environment, The Thai Government. 1992. *Thailand Country Report to the United Nations Conference on Environment and Development.* Office of the National Environment Board, Bangkok.

Thailand Development Research Institute TDRI. 1987. *Productivity Changes and International Competitiveness of Thai Industries.* Thailand Development Research Institute. Bangkok.

Chapter 7

Problems of Ecosystem Devastation as a Focus in Environmental Education

Tatuo Kira, International Lake Environment Committee Foundation, 1091 Oroshimo-cho, Kusatsu City, Shiga Prefecture 525-0001, Japan.

The ILEC Environmental Education Project

The International Lake Environment Committee (ILEC) implemented a seven-year pilot project for promoting environmental education in primary and secondary schools in six countries with widely different natural, cultural, and socioeconomic backgrounds. Its basic approach was: 1) to take a nearby lake, reservoir or river and its watershed as the material for teaching; and 2) to start the course with recognition of, and direct contacts with local problems of aquatic environments. This approach fulfilled the needs mentioned above, proved very successful, and is expected to provide a model for further environmental education.

The International Lake Environment Committee (ILEC) is a non profit foundation based in Kusatsu City, Shiga, Japan, established in 1986 with the objective of contributing to the conservation of environments in the lakes and reservoirs of the world and the sustainable use of their resources. One of the main undertakings so far made by ILEC was a seven-year pilot project for the promotion of environmental education (EE) in primary and secondary schools in six countries: Argentina, Brazil, Denmark, Ghana, Japan and Thailand.

The main objectives of this project were:
1) to give students a correct understanding of the causes and processes of environmental disruption in inland waters at appropriate levels for their grade; and
2) to bring students directly into contact with an actual environmental problem in their immediate surroundings and to let them experience it in order to help their study and enhance their willingness to cope with the problem.

Two Categories of Environmental Problems

Aside from the details of the project which will be discussed later, it must be borne in mind that there are two groups of environmental problems which are different from each other with respect to their causes, the processes of their occurrence, and the necessary countermeasures.

One is the group of problems originating from our industrial activities or industrial products (here the word *industry* is used in its narrow sense to cover secondary industries in general) such as air pollution, acid precipitation, contamination of water and soil with man-made chemicals, degradation of water quality by eutrophication, etc. These were originally problems in industrialized and urban areas, but their impact has now been globalized, causing global warming, ozone layer depletion and other worldwide problems.

The other group, on the other hand, includes the problems resulting mainly from excessive or unreasonable activities in primary industries, such as deforestation, desertification, accelerated soil erosion, and the degradation of semi-natural ecosystems of agricultural and pastoral lands, fishing sites in inland/coastal waters, etc. Those ecological disasters are the most serious and frequent environmental problems in the rural areas of the world, where population pressure and the thirst for a better standard of living among local communities are causing imprudent exploitation and over-use of their lands and waters. Ecological disasters need specific countermeasures for their solution different from those for the former group of problems. Environmental education must inevitably aim at improving people's knowledge and understanding, not only of the problems of the former category, but also of those of the latter.

Ecological Disasters

However, this latter aspect tends to be ignored in school environmental education, especially in industrialized countries where people are keen about environmental education but have little chance of direct contact with ecological disasters. In regions suffering from ecological disasters, on the other hand, environmental education has not yet become widespread. I should, therefore, like to stress in this paper the necessity of dealing with ecological disasters as one of the most urgent themes in environmental education.

Some of the ecological disasters currently taking place, such as excessive exploitation of tropical forests, are now recognized as a matter of major environmental concern by the global society, but it should also be noted that there are other types of ecological disasters which are not as visible as the destruction of primeval forests but are as important or have even more serious implications for the lives of local peoples.

Degradation of the Resource Bases for Primary Industries

Desertification is an example of this process. It is not the devastation of undisturbed natural landscape. Sparse scrubs and grasslands fringing desert areas are mostly semi-natural vegetation which has long been utilized for grazing and fuel collection by local inhabitants. If population increase and other socio-economic factors lead to the more intensive use of such semi-natural ecosystems, the plant cover is successively lost, soil erosion by wind and rain water then starts, and eventually the whole of the land is left denuded. The climate of arid zones is characterized by the periodical alternation of dry and wet years. During a spell of wet years, people cultivate grasslands for crop production, but the land which has lost its original grass cover is subject to severe wind erosion once the climate enters the next dry cycle. The tragedy caused in this way in the middle west of the United States in the 1930s still survives in people's memories in its name, "Black Sunday" (April 14th, 1934), when the first terrible sandstorm struck the region.

The degradation of agro-pastoral systems is not limited to arid regions. In communities subsisting on agricultural production, population pressure is increasing the needs for food and fuel while the diffusion of the worldwide market economy is stimulating people's desire for more cash income to buy modern commodities for a better standard of living. As a result, lands tend to be more and more heavily used as people cultivate unsuitable sites (Figure 1), plant cash crops which are less tolerant of soil erosion (Figure 2), shorten the fallow period in the crop rotation cycle, increase the number of grazing animals per unit area of pasture (Figure 3), and so forth. All those cases of land overuse accelerate the loss of fertile surface soil by erosion, depriving the land of fertility semi-permanently.

The Tanakami Mountains, which can be seen to the east of Shiga University across the Seta River, were subject to similar devastation around the beginning of this century. The picture in Figure 4a shows a view of the mountains in 1914, when the slopes were almost bare and treeless. This area had been a common ground for collecting plant biomass and litter to be used as compost and fuel by nearby villagers, but it lost its plant cover due to excessive harvest during the 18th and 19th Centuries. Slope protection work by means of terracing and planting of grasses and shrubs had just started at that time, and has continued up to the present with enormous investments of money and labor. Eighty years of effort have allowed the stands of pine to recover to the extent shown in the bottom photograph in

Figure 1. Rural Landscape on Yunnan highland of China (Dali District, Yunnan Province) in the upper reaches of the Yangtze River. Hill slopes are extensively cultivate without terracing. Sparce stands of pine in the background are used for grazing cattle and buffalo.

Figure 4. This case illustrates how difficult it is to restore ecosystems once they are devastated.

Sites of primary economic activity in former days, such as fields, grazing lands, and fuel-producing forests for firewood and charcoal represented semi-natural ecosystems in which human actions were dynamically balanced with the functions of the natural environment to ensure sustainable production. Those traditional systems are, however, being degraded and losing their production capacity rapidly and semi-permanently in the current world economic situation. This is really a serious crisis in view of the expected situation in the next century when the world population will be doubled, endangering food supplies. Restructuring the world's primary economic activities and restoring their ecological resource bases are, therefore, among the most pressing needs for the whole world. Recognition of the seriousness of these ecological disasters should be among the main targets of environmental education.

Problems of Ecosystem Devastation

Figure 2. Growing cash crops (tobacco in case) often accelerates soil erosion. Dali District, Yunnan, China.

Figure 3. A pine forest on Yunnan highland where undergrowth has been lost due to overgrazing and is subject to severe soil erosion and landslide. Dali District, Yunnan, China.

136   Integrated Environmental Management

Figure 4. Tanakami Mountains (Otsu City, Shiga, Japan) in 1914 (top) and 1995 (bottom). Slope protection work could bravely recover the poor pine forests in more than a half century.

The Case of Aquatic Ecosystems

Degradation is also going on in aquatic ecosystems such as rivers, lakes and coastal sea waters which have been used for fisheries and a sources of water, though the process tends to be less noticeable than in terrestrial cases. As regards the case of lakes, with which ILEC is mainly concerned, environmental degradation is really critical given that the lakes' contribution to our lives is growing all the more important. Fishery products from lakes are still an important protein source for inland rural communities, while the supply of freshwater from natural lakes and reservoirs (man-made lakes) is steadily increasing, even in large cities. As shown in Figure 5, however, lakes that used to continuously supply us with water and fishery products are now suffering from rapid sedimentation caused by accelerated soil erosion in catchment basins, water level decline due to excessive water withdrawal, eutrophication caused by the inflow of nutrient-rich waste water, acidification of water due to acid precipitation resulting from air pollution, and contamination with man-made toxic substances. All these processes cause the decline of water quality and quantity, leading to the loss of the lakes' role as a water source, the collapse of fisheries, and the loss of biodiversity through the extinction of aquatic organisms including a number of endemic species.

Lake Dongting, the second largest freshwater lake of China, is a big retarding basin for the Chiang Jiang (Yangtze River) connected with the midstream of the river. Flood water during the rainy season flows into the lake through a network of watercourses, staying there for some time leaving a large amount of sediments on the lake bottom, and flows out again to the Chiang Jiang. The sedimentation rate amounts to several centimeters per year, because the river water is now fully loaded with soil owing to the overuse of watershed lands as shown in Figures 1-3.

Dongting was the largest freshwater lake in China until the 1950s, but its water surface has now decreased to about 60% of the former area, and this is restricting boat transportation on the lake which is extremely important for industries and communities around the shore. If the high sedimentation rate continues into the future, the lake will almost disappear within the next century. When the giant Chiang Jiang Dam now under construction some distance upstream from Lake Dongting is completed the sedimentation problem will be solved, but sediment loads will then seriously shorten the dam's life. Similar sedimentation problems are common in a great number of smaller lakes throughout the world.

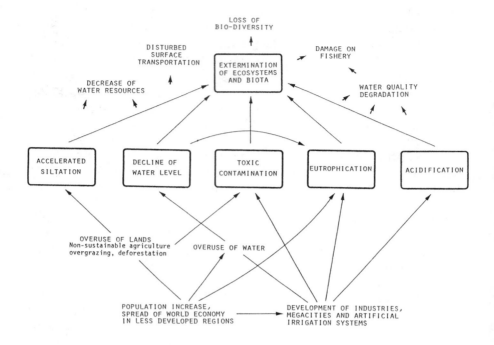

Figure 5. Six major environmental problems in world lakes and reservoirs (Kira, 1993).

The tragic shrinkage of the Aral Sea, the world's fourth largest lake in the desert area of central Asia, caused by the extensive development of irrigated farmlands along its inflowing rivers is now well known. Owing to this development project initiated by USSR government in the 1960s, the diversion of water from two major tributary rivers limited the amounts reaching the lake, so that the evaporation from the lake surface, which had once been balanced with the water inflow, reduced the lake's original surface area (100 times as large as that of Lake Biwa) by about 60% and the water volume by about 80% until the 1990s. The salt concentration of the lake water increased from 28% to 100% of that of sea water, all native organisms in the lake were exterminated, and the coastal people were deprived of fishery products. The dried-up lake bottom, up to 100 km in width, turned into a salt desert with frequent sandstorms causing damage on crop fields, while the rising salt content of drinking water is a serious hazard to human health. Many other lakes in the arid zones of Africa, Near East and western United States are also suffering from a similar situation.

Eutrophication is the most frequent and troublesome type of environmental degradation in world lakes. It may be said that almost all lakes in densely populated and/or industrialized areas are more or less suffering from eutrophication. They are also often suffering the effects of toxic contamination. Where sewage treatment and other measures for reducing nutrient loads in lakes have not yet been taken, lake water tends to become eutrophic to the extent that the water surface is frequently covered with blooms of blue-green algae. Under such *hypereutrophic* conditions, the water becomes anoxic in the deeper layers killing fish and benthic animals, and greatly impoverishing the lake's biota. These lakes can hardly serve as sources of tap water. Lake Biwa has also been subject to eutrophication since the 1960s. Despite the efforts to stop its progress in order to assure the supply of good quality water to a population of some 14 million people in Shiga, Kyoto, Osaka and Kobe, algal blooms have been spreading for the last 20 years. The restoration of water quality of eutrophic lakes is so difficult both technically and financially that eutrophication is really a great threat to the future freshwater supply for the increasing world population. It is even anticipated that the shortage of freshwater may become a global issue earlier than food shortage in the near future.

The Incorporation of Ecological Principles into Environmental Education

The various abnormal processes in ecosystems that emerge as ecological disasters are the result of highly complicated causes. In order to understand and cope with an ecological disaster, one has to understand, to some extent, not only about the complex interactions among many kinds of organisms and diverse environmental factors, but also about the interrelations between the causal factors, both ecological and socio-economic. Of course this also applies to the study of environmental problems of industrial origin, but wider knowledge and a deeper understanding of ecological processes are evidently needed in the case of ecological disasters.

However, it is a very difficult task for school pupils and the general public who lack the basic knowledge to understand the complex background to environmental issues. Indoor lectures alone can hardly achieve the goal. Therefore, the ILEC EE Project has adopted the strategy mentioned in the objectives listed above. At the start of the EE course, students are taken to observe and experience an environmental problem near their schools. The students may be able to draw an inference intuitively on why the problem has occurred and which factors are responsible for its persistence, because they are naturally well acquainted with the surrounding area. The environmental problem is directly related to their daily life in most cases, so that they are expected to be greatly concerned about it and very interested in observing and discussing it.

After the students have benefited from these experiences, they will be able to acquire basic knowledge more easily in the class room. When the students ask new questions, they can make repeat visits to the site of the problem and, if necessary, carry out laboratory experiments to answer the questions. Through these steps, they will be well prepared to consider what should be done to solve or mitigate the problem. This approach may be also useful for other subjects, but its advantages seem particularly great for environmental education.

As for ecological disasters, it is important to make students realize the connection between relatively simple ecological principles and what is actually taking place in their environments according to this approach. They have to build capability of finding signs of something unusual in the environment which might grow into a serious problem if left unattended.

The Lake/River Watershed Approach

One may wonder whether teachers can always find such suitable sites for environment oriented experience for their students. The following provides an answer to this question.

The uniqueness of the ILEC EE Project was that a river or a lake near the schools, together with the whole area of its watershed, was taken as the site for environmental education.

Rainwater from the sky passes throughout the entire watershed like the blood in animal bodies, nourishing natural and agricultural ecosystems, supporting the lives of residents, and eventually flowing into a river or a lake carrying with it a load of various substances. Therefore, a river or lake and its watershed area can be regarded as a unitary system bound together closely by the flow of water. The quantity and quality of water as well as the flora and fauna in rivers and lakes respond sensitively to the natural and anthropogenic changes in watershed environments. Lake and river watershed systems with such properties offer excellent material for studying the functions of natural ecosystems and the effect of human activity on the environment. The impact of human life on water quality and aquatic life can be easily observed everywhere where a settlement and a stream are present.

The ILEC EE Project was carried out on the basis of such a common methodology in six countries in which natural, socioeconomic, and cultural conditions were widely different, and it proved that this watershed approach could be easily and effectively applied in each of the countries. The outcome of the project has recently been published in one of the ILEC Guideline Book Series (Jørgensen, et al., 1997).

The most effective way to improve environmental education is to base it firmly on the local environment where the instruction takes place. This affords a strong incentive to improve local environments, and to develop the educational experience into the mass effort required to safeguard our globe. This was the conclusion brought about by the ILEC EE Project.

References

Jørgensen, S. E., M. Kawashima, and T. Kira, Eds. 1998. *A Focus on Lakes/Rivers in Environmental Education (Guidelines of Lake Management)*. Kusatsu, Japan: Environment Agency, Government of Japan and International Lake Environment Committee Foundation, p168.

Kira, T. 1993. Major environmental problems in world lakes. *Memoir dell'Istituto Italiano di Idrobiologia*, 52: 1-7.

Chapter 8

Education for Solving Environmental Problems: How to Develop Teaching Materials and Generate Support for Environmental Education in Schools

Munetsugu Kawashima, Faculty of Education, Shiga University, 2-5-1 Hiratsu, Otsu, Shiga Prefecture 520-0862, Japan.

The Urgent Necessity and Importance of Environmental Education

These days, human beings are faced with many global environmental problems such as acid rain, ozone depletion, global warming, and ocean pollution, as well as more localized environmental problems. Although the development of science and technology has brought convenience and wealth, particularly in developed countries, a society of mass production and consumption has caused environmental destruction. The solution to many environmental problems depends on whether we can control new technologies which are the by-product of human development.

On the other hand, the concept and importance of environmental education have been discussed and developed worldwide over the last three decades. Since the United Nations Conference on the Human Environment in Stockholm in 1972, many conferences have been held to address environmental problems. The importance of environmental education has also been emphasized since the International Environmental Education Workshop held in Belgrade in 1975. The Belgrade Charter defines clearly the goals of environmental action and education:

Goal of Environmental Action: To improve all ecological relationships, including the relationship of humanity with nature and people with each other.

Goal of Environmental Education: To develop a world population that is aware of, and concerned about, the environment and its associated problems, and which has the knowledge, skill, attitudes, motivations and commitment to work individually and collectively toward solutions of current problems and the prevention of new ones.

More than 20 years have passed since the Belgrade Charter defined and circulated these excellent objectives of environmental education. However, these goals have not been achieved. Meanwhile, especially in industrialized countries, education has mainly supported the development of science and technology which has brought us mass production and mass consumption. Environmental problems are reaching critical levels and are becoming more difficult to solve. The importance of education to solve them should be emphasized all over the world.

Brief History of Environmental Education in Japan

Environmental Education in Japan had its origin in anti-pollution education. Since 1950, human health hazards caused by environmental pollution, including Minamata Disease and Itai-itai Disease, have occurred frequently: Japan has sometimes been called an "advanced nation in pollution." In 1964 a national research organization for pollution control started investigations into the origin and degree of injury due to pollution among elementary and junior high school students.

The Environment Agency was established in 1971. In the same year, the course of study for schools was partly revised by the Ministry of Education according to the revised "Basic Law for Environmental Pollution Control" (1967). The national guidelines for Social Studies clearly stated that "it was very important to protect people's health and life from industrial pollution."

Since the beginning of the 1970s, groups of conservationists have become more conscious of the need for education to protect natural environments. The courses of study were revised for elementary and junior high schools (both compulsory education) in 1977 and for high schools in 1978 to incorporate the study of environments and natural resources in almost all subjects. Furthermore, a new course was designed by the Ministry of Education in 1989, adopting the new subject, "Life Studies," in the first and second grades of elementary school. The contents of such subjects as Science and Social Studies were more strongly related to environmental issues in all grades in elementary and junior high schools. In 1991 the Ministry of Education announced guidelines for environmental education in which global environmental problems such as acid rain, global warming, and water pollution must be dealt with in relevant subjects. The new course of study was started in April, 1992, in elementary schools, 1993 in junior high schools, and 1994 in high schools. All the schools were obliged to adopt the new course of study. It will be revised after a few years, then all schools will be required to introduce environmental education in an integrated manner. Thus, environmental education is steadily advancing in Japan, but there is a need to develop teaching materials and curricula.

Lifelong Learning and Environmental Education

Environmental education is extremely important for adults as well as for children, and it is regarded as a significant theme in lifelong learning. The adults have destroyed the environment, so solving environmental problems is a big challenge for them. However, the warnings that should be directed to adults are often given to children. It is often pointed out that children today are not exposed to the natural environment; therefore, they have lost recognition, love, and sensitivity toward nature. This tends to be regarded as a "children's problem." However, it is the society created by adults that has deprived children of the opportunity to have direct contact with nature and has compelled them to adapt to the deformed modern society with technological development. Therefore, it is necessary for adults to reflect on this fact and to think about the global and regional environment together with children.

The state of the environment should not be judged mainly by scientists or administrators, but by citizens. Citizens should play a major role in overcoming the environmental crisis. Therefore, it is essential that citizens have the ability to:
1) understand environmental problems with scientific eyes;
2) make decisions about technology and society; and
3) act on the basis of their own decisions.

A reorientation in lifestyle will change the present socio-economic system (production, distribution, consumption, disposal, and others) towards an environmentally sound society. Consequently, the preservation of nature and the solution of environmental problems are important objectives for all human beings as a social subject in lifelong learning. The importance of environmental education for the citizen is recognized and emphasized.

However, many people seem to be little concerned with environmental problems and do not know how to protect environments, simply because they have had no chance to learn about the environment during their school and college days. Today, some symposia and lectures about environmental problems are open to all citizens, and this will contribute to the promotion of people's awareness about environmental improvements. Furthermore, environmental education in schools should be emphasized in addition to the importance of lifelong learning because school education is now recognized as the basis of lifelong learning.

The Development of Teaching Materials

One of the reasons why environmental education has not been sufficiently implemented seems to be that its methodology has not been developed although the objective have been well defined. Curricula and teaching materials should be developed both in schools and in-service education.

Environmental problems can be dealt with in the framework of existing subjects in school education, but some aspects may not be suitable for science or social studies, so different perspectives for teaching materials may be required. The following points should be considered in this context:

1) The causes and effects of many environmental problems are not clearly identified. Besides, all environmental problems are ongoing. It is important to recognize where education becomes necessary to address these problems without the causes or effects being fully identified.
2) It is important to convey the correct information about environmental problems to students, but we should realize that this is not always possible. In addition, the future perspective of the problem should be discussed, but it is also important to keep in mind that scientific predictions are uncertain.
3) Environmental problems do not arise over a short period of time, but develop slowly. We need to take the element of time into consideration and not be hasty in attempting to describe the causes of environmental damage. It is necessary to observe ongoing changes with students participating in experiments, research, and other observations and learning experiences.
4) Environmental problems are not remote from the experience of children; they are all around us. Therefore, it is important to conceive of them as problems adjacent to the social structure or system where we live. It is advisable to conduct as many studies as possible that utilize local environments. It is also important to put in the context of local or daily life the teaching materials for environmental education, giving consideration to the characteristics of learners in different grades. Adopting daily life materials or local materials makes it possible to develop studies through observations and experiments.

5) Environmental problems are often discussed from the natural science point of view. This means that the destruction of nature or the ecosystem tends to be emphasized. Because environmental disruption was brought about by human activities, it has both social and scientific background. This is an important aspect that must be considered in promoting environmental education.
6) At the end of the course, students usually are provided with a summary of it which contains more than what they have covered, which often confuses them. It is recommended to summarize only what they have learned and give them enough time for discussion. It is also important to develop teaching materials that answer their questions.

Based on the above discussion, the following points important for EE are summarized:
1) We should set a certain location for the studies based on the students' own experience and give them as many opportunities as possible to come in contact with natural objects and phenomena and to observe what human society really is today.
2) We should provide as many direct experiences as possible by the use of experiments, observations, and research to nurture an inquiring attitude. A scientific view and way of thinking should be emphasized.
3) We should aim at studies and activities that are continuously enjoyable.

Consequently, it is essential that teachers understand the environmental problems near the schools and design environmental education programs that satisfy their students' desires to learn about the environment in depth. Therefore, it is imperative that teachers develop a positive and voluntary attitude to create teaching materials. In elementary and junior high schools, it is important to use materials for environmental education from the local area or in the place where the children live as long as they are suitable to their developmental stage. Therefore, it is desirable to develop the environmental education activities and lessons with daily life materials and relevant experiences. Environmental education in the future should not be regarded as merely outdoor or natural education, but it should be developed as a creative activity that will enhance the quality of the earth and the regions where natural disruption and environmental aggravation have progressed. This will create a more desirable environment.

Environmental Education Classes

Many environmental classes are still focused on conveying knowledge one-way to the students. This may also be applied to many other subjects. However, we should move from conventional education which emphasizes common knowledge and skill into education which develops the ability to think, judge, and act. This is especially important for environmental education. We have studied the environmental education program based on the students' experience, in order to nurture a scientific view as well as the power of execution which leads to the solution of environmental problems, the ultimate objective of environmental education. To realize this, we have concluded that, particularly in elementary and junior high schools, it is indispensable to create an environmental education program that satisfies the students' desire to understand the environmental circumstances and problems in the neighborhood of their school and to know more about the environment. In addition, we would like to adopt active discussions and role-playing conducted by the students themselves. After these activities, positively introducing learning by experience of environmental problems and discussions of the content of what they learned, the way to solve environmental problems, what they can do now, or what they will be able to do when they grow up, will greatly contribute to fostering the power of judgment.

At present, many countries do not teach environmental education as an independent subject, nor do they provide systematic guidelines on its content. Because the daily classes at school are normally based on the course guides/outlines prepared by the government, there are few chances for environmental education to find a niche. Therefore, the relationship between each subject and environmental education is important. The 1994 course of study/guide provided by the Ministry of Education in Japan suggests that there must be a connection between different school subjects and environmental education. "It is desirable to conduct environmental education in every subject at school, however, the treatment of the phenomenon related to the environment should naturally be considered in accordance with the characteristics and objectives of each subject. In this case, the point to be considered is linking the phenomenon with each subject in an appropriate manner." In short, it is possible to conduct environmental education as overall, systematic study within the time deemed by the school as necessary. Frankly, though the importance of environmental education is emphasized, the time allocated to this purpose is very limited. Also, schools are obliged to depend on the level of each teacher's interest or competence. As a result, many environmental education classes tend to be a "thrown-in type" centered on just explaining environmental problems during classes on a different subject, with teachers the only ones providing knowledge to the students. Because environmental education is related to the other subjects, the wrong impression is sometimes given that it can be dealt with as part of any subject, and with any kind of material. However, we should keep in mind that science and environmental education stand on two essentially different foundations. Science is based on natural science where natural principles and natural laws are clarified. Environmental education is based on environmental science which studies the methodology to perceive and solve environmental problems. Environmental education is, therefore, also linked closely to social science. Also, it is necessary to realize that environmental education is not a mere study of the environment but a study for the environment. In this respect, it is expected that environmental studies, or something like it, will be introduced as a separate subject in the future.

Under the current system, we are required to carry out environmental education with other subjects. We then need to develop a program which can simultaneously achieve the objectives of each subject and environmental education. Furthermore, because the available class time is limited, it is important to set fixed objectives.

Support Systems for Environmental Education: The Role of the Center for Environmental Education and Lake Science, Shiga University

In view of the necessity and importance of environmental education in school, all teachers who have direct contact with the younger generation should make every effort to develop pertinent methods and teaching materials for environmental education. However, there are still many teachers who are less concerned with environmental problems and education, simply because they had little chance to learn about the environment and its problems during their school and college days. The present situation points to the fact that environmental education is not an independent subject in elementary and secondary schools and the methodology of environmental education is not treated systematically in the courses of the faculties of education or as part of in-service education. In this respect, it is necessary to design a system to help teachers study and upgrade their skills in environmental education.

In order to study and expand environmental education, the Center for Environmental Education and Lake Science was established in April, 1995, as an attached institute of the Faculty of Liberal Arts and Education, Shiga University. The main purposes are:
1) to implement the interdisciplinary and collective study of the environment by using Lake Biwa and its watershed, the agricultural and forest areas associated with the university; and
2) to train coordinators for environmental education and expose them to opportunities to gain experience which will enhance their practical ability and theoretical knowledge in relation to the environment.

The Center is composed of full-time staff and guest researchers who are interested in environmental education. The full-time staff started to offer lectures about environmental education to all of the students of the Faculty of Liberal Arts and Education, Shiga University, and opened the course of environmental education for citizens in partnership with the Shiga Prefectural Board of Education. In addition to pre- and in-service education, in 1996 the Center started "The Shiga Project for Lake Environmental Education" in which schoolteachers and citizens participate freely.

The Shiga Project for Environmental Education

The purposes of the project are: 1) Support of environmental education in schools; 2) Development of teaching materials on local environments, and 3) Exchange of information. Through discussion, the Research Committee of the Project planned a participatory program named "Let's all make maps of aquatic environments together." This program was tried twice in the summers of 1996 and 1997, focusing on water qualities of the rivers in the watershed of Lake Biwa. Participants took the water samples from the rivers near their schools or houses at almost the same time on the 22nd of August and then brought them to the Center. In 1996, there were 91 participants and 133 samples, and in 1997 there were 80 participants and 186 samples. More than 80 man days were needed to analyze the water samples. The following items were measured and mapped: electro conductivity, turbidity, pH, COD, total phosphorus, $H_2PO_4^-$, total nitrogen, NH_4^+, NO_3^- and major ions such as Na^+, K^+, Mg^{2+}, Ca^{2+}, Cl^- and SO_4^{2-}. The maps are available from the home page of the Center <http://www.sue.shiga-u.ac.jp/WWW/kosyo/kosyo.htm>. These maps are useful to understand the present situation of rivers in the watershed of Lake Biwa. This program also seemed to play a role as a means of encouraging teachers' and citizens' awareness of local environments. Some maps were used in environmental education classes in schools. The aims of this program are: 1) to support environmental classes in school, 2) to study environment improvement based on the local environment, 3) to support grassroots movements, and 4) to solve environmental problems, especially the eutrophication of Lake Biwa. In order to attain these aims, it is expected that teaching materials will be developed based on the maps including vegetation, animals, and landscapes as well as the water quality of rivers in the watershed of Lake Biwa. The fundamental study of factors that influence aquatic environments is also necessary from the viewpoint of land use, history, and human life.

Support for Refresher Education of Teachers in Thailand

Since 1995, the Center has supported the development of teaching materials for environmental (science) education and the study of the methodology for the promotion of environmental education in Thailand, which has many severe environmental problems. In Thailand, teachers' awareness of environmental problems is not high. Even if they are concerned about environmental education, they have little experience of environmental education based on students' learning experiences. A joint study with Chiang Mai and Prince of Songkla Universities has been initiated from the viewpoint of developing teaching materials and educating teachers on the importance of environmental education. A short training course on environmental education for teachers was conducted twice in Chiang Mai and once in Pattani. The course stressed not only lectures but also the observation of natural environments and the investigation of problems, especially water quality of the rivers, the garbage problem, air pollution and the destruction of mangrove forest. We were able to help teachers monitor these using simple analytical instruments. Through these three training courses, we were able to demonstrate the applications of environmental education and raise teachers' motivation to solve environmental problems. This type of training course is considered to be very effective and the support for field work is very important in environmental education. As one of many results, a teachers' book of guidelines for environmental education was written in Thai and this is expected to be useful for the promotion of environmental education.

Development of an Information Exchange System

Many environmental education programs have been developed independently, but the lack of an information and exchange system among schools and/or citizens should be addressed. It is also important to establish an international network on environmental education. To accomplish this, it is necessary for a number of environmental education institutes to take the lead in the exchange of information on educational programs between schools. As one of these pioneer institutes, the Center for Environmental Education and Lake Science at Shiga University will play an important role. It plans to create a database of teaching materials and programs on environmental education and offer these for use in schools.

Chapter 9

The Present Situation and Problems of Environmental Education in Shiga Prefecture

Shigeki Shirai, Board of Education, Shiga Prefecture, 4-1-1 Kyomachi, Otsu, Shiga Prefecture 520-8577, Japan.

Introduction
The History of Environmental Education in Shiga

Between the mid-1950s and the mid-1960s, many pollution problems arose nationwide, including the increase of waste water pouring into Lake Biwa. As a result, the amount of nutrients like nitrogen and phosphorus compounds rose markedly. This caused phytoplankton to increase dramatically, resulting in the deterioration of water quality through eutrophication. This in turn resulted in a foul smell in the drinking water in 1969, and the appearance of "red tides" in 1977, and "Aoko," or green bloom in 1983.

In Shiga, education programs relating to public health and natural conservation were gradually transformed into programs in environmental education. This was in order to take into account the impact of human existence and lifestyles on the environment, which the old programs did not cover sufficiently. However, many people in the educational system failed to understand the meaning of "environmental education," and for a while from the late 1960s on it was taken to mean "renovation of the educational environment," even though real programs in environmental education had already begun. Administrative measures for the promotion of environmental education improved the situation in allowing the subject to spread and develop.

The Prefectural Board of Education published a *Handbook for Environmental Preservation* (1974), and *Collections of Good Activities for Environmental Education* (1976). Starting in 1980, the Prefectural Board of Education selected thirty schools for the development of environmental education. During the 17-year program these have included schools in most towns and cities in the prefecture, with the result that the nature of environmental education is now understood by all teachers and students. The Board also published a text book on environmental education at the same time, to assist teachers. In 1981, the Prefecture held a conference on environmental education for primary school teachers. It also implemented an ordinance concerning the prevention of eutrophication in Lake Biwa in 1980, establishing the Lake Biwa Research Institute in 1982, launching the fieldwork vessel "Uminoko" as a floating school in 1983, and organizing the first International Conference on the Conservation and Management of Lakes in 1984. Big projects such as these have made people concerned about the environment, and environmental education has spread to every school.

Environmental promotion and environmental education in Shiga

The prefectural government has adopted as its basic principle for prefectural administration what it describes as the 'creation of 'New Omi Culture'." One of the main concepts within this is "living with nature," on which the government has developed its environmental administration. In 1996, Shiga implemented its basic Ordinance on the Environment. Many environmental projects under the Special Measures Act for the Lake Biwa Comprehensive Development Project were completed in 1997. In 1998 the relevant divisions of the prefectural government were reorganized as environmental authorities. Following the Lake Biwa Comprehensive Development Project, in-depth environmental protection policies are being developed. We are also now drawing up a comprehensive plan which will enable us to pass on a clean and productive environment to the next generation. Environmental education will become part of this and will be promoted more and more in the future. It is currently being promoted by the School Education Division, the Eco-Life Promotion Division, and the Life-Long Education Division which organize the Omi Continuing Education College within the Prefectural Board of Education. Other programs on the environment are promoted by nine divisions belonging to the Lake Biwa Environment Authorities. In 1998 environmental supervisors were appointed in each division and in the Prefectural Board of Education the Vice-Chief of Education serves in that role.

Schools in Shiga

In Shiga, there are 234 public elementary schools, 98 public junior high schools, and 64 public high schools, in addition to two elementary schools, three junior high schools, and four high schools operated privately. Programs of environmental education are in force in each public, national, and private school. This section discusses the current situation and the main problems of environmental education in the Shiga Prefecture school system.

The Present Situation of Environmental Education in Schools

It can be reported that there has been an improvement in many aspects related to environmental education, such as teaching methods and techniques, the issue of textbooks, the promotion of study, the provision of information, the design and execution of the curriculum, and the number of activities taking place.

Improvement of teaching methods and techniques

The key points for the promotion of children's environmental education are what to learn and how to teach it. Teachers' understanding and improved teaching methods and techniques are necessary as environmental education needs to be carefully integrated into the curriculum. In-service training is, therefore, provided by the School Education Division and Comprehensive Education Center of the Prefectural Board of Education to strengthen the teachers' knowledge base.

1) *Study Conference on Environmental Education.* This study conference is held each year for teachers in charge of environmental education in primary, and secondary schools, as well as special schools for handicapped children. It includes discussion of the introduction of practical activities in each kind of school together with lectures.

2) *Training in environmental education for newly-appointed and experienced teachers.* In order to offer all teachers training in environmental education, in-service training is provided for all newly-appointed teachers and for teachers in their sixth year of teaching.

3) *Environmental Education Seminar (Comprehensive Education Center).* Teachers can undergo voluntary training in environmental teaching methods and techniques including research methods, as well as visiting environment-related facilities.

Textbooks on environmental education

Recognizing the need for teaching materials, the Prefectural Board of Education has issued textbooks on environmental education for each level of the school system from 1980. The books are designed to help every teacher in every school, and not only the teachers

specifically concerned with environmental education. These textbooks have the following features:
1) The elementary school edition includes a workbook for pupils and a handbook for teachers. The workbook is sent to all fourth grade pupils every year for them to use in the three years until their graduation.
2) The junior high school edition contains a file of materials which is sent to every teacher to help them create worksheets using their own ideas.
3) These materials can be used in helping children learn through experience in science, social studies, homemaking, moral education, and special curricular activities, and the sheets in the file contain teaching plans as models.
4) The textbooks contain discussion of global environmental problems, including issues relating to Shiga Prefecture such as the aquatic environment of Lake Biwa.
5) Each edition is revised every five years to update the data, and the editions currently in use are the third edition for elementary schools and the fourth edition for junior and senior high schools.
6) Each textbook has 120-130 pages printed in color, and schools are supplied with copies of each new edition.

Promotion of the study of environmental education

In order to promote and develop environmental education, we need both for it to be practiced widely in every school and to make further suggestions as to how it can be used. The Prefectural Board of Education has the following programs to promote it.
1) *Appointed Schools for the Study of Environmental Education.* The Prefectural Board of Education appoints one elementary and junior high school each year, to study improvements to the curriculum in order to improve guidelines for education on environmental issues. The schools selected report their results after two years of research.
2) *Model Schools for Environmental Education.* Each school practices a form of environmental education which takes into consideration the characteristics of each area and its children. We have established a system of model schools which can provide support for this education and enrich it by making available new ideas to all the schools in Shiga. This model school program began in 1996, replacing the previous program involving thirty schools which began in 1980. Almost every

town and city has had experience of the previous program and environmental education has spread throughout the prefecture. We now face the problem of improving teaching methods and content. The model schools were, therefore, selected from among the schools which had developed distinctive programs in environmental education. We have selected thirty model schools in 1998 to act as associate research institutions of the Lake Environment Training Center of Shiga University. The Center will gather information on environmental education from them and provide them with support. The model schools hold a study conference twice a year.

Information provided on environmental education
Teachers often teach without outside monitoring, and environmental education is no exception. They, therefore, need to examine what forms of education are most suitable and the areas in which there is room for improvement. The materials provided are also guides to putting environmental education into practice, and they should be available in every school in Shiga. The Prefectural Board of Education regularly issues a collection of reports from the schools involved in its selective programs, and sends it out to all the schools and authorities concerned. The eleventh edition has been published.

The design and implementation of the curriculum
Environmental education carries over into all other subjects and must be fully integrated. In each school, therefore, we either introduce environmental issues into other subject units, or we plan original integrated study programs within the time allocated by the school. Schools and teachers have different ideas for carrying out environmental education. Every school has a general plan to establish environmental education and a teaching plan to put it to action. Through this steady effort there remain only a few schools not practicing environmental education. In the present six-day week curriculum, two Saturdays a month are holidays. Generally schools assign teaching periods to other subjects, and it is very difficult to practice environmental education within the regular timetable. To find a way out of this difficulty, teachers must improve their teaching techniques and teach interdisciplinary courses together regularly. The Central Educational Council expects teachers to teach periods of integrated studies.

Distinctive activities in environmental education in Shiga
1) *The Lake Biwa floating school.* Every pupil in the fifth grade of elementary school has a two-day study sailing trip on board the ship Uminoko, which is 65 m long, weighs 928 tons, and has a seating capacity of 240. Studies which can only be carried out on the lake are very important. Pupils can view Lake Biwa from the shore, but few of them normally have the opportunity to go out on to the lake. This two-day study sailing trip is a golden opportunity for them both to feel their bond with Lake Biwa and to understand its environmental problems through direct experience. This type of experiential learning is very valuable for children in elementary school. It is regarded as valuable by the pupils and their parents, and is also a matter of pride for the prefecture. We made 93 study trips last year. The ship was commissioned fifteen years ago, and up to now 250,000 pupils have been on board. This is about one-fifth of the population of Shiga Prefecture, and the proportion increases year by year. It is important for the people of the prefecture to feel attached to Lake Biwa and to be aware of its value, and it is, therefore, necessary to continue the valuable environmental education which the ship provides.
2) *Development of the Program of Environmental Education Through Experiential Learning.* In order to tackle the root of the current educational problems, it is necessary to rethink the idea that education is merely knowledge and data, and to begin to nurture creative talent, judgment, self-expression, and self-realization. Experiential learning is a useful way of reflecting on one's way of life and existence, and so we have had a program running since 1966 to promote learning through experience in subject studies. There are five features of this program. One of them is the relationship with nature, and environmental education is practiced with this in mind. It forms the basis of 45% of the programs registered in schools. Teachers reinforce children's consciousness of various environmental processes through topics such as breeding experiments, cultivation, the recovery of resources, and cleaning up the environment in order to exchange passive activities for those based on direct experience which are most valuable in children's education.
3) *Lake Biwa Day and environmental preservation.* The Biwa Eutrophication Ordinance, the Landscape Protection Ordinance, and other ordinances concerning the environment of Shiga Prefecture came into effect on July 1st, 1984. Therefore, July 1st

has been designated Lake Biwa Day, and environmental preservation activities are held in every part of Shiga Prefecture to emphasize the importance of our environment. Every school mounts involving the local citizens.

Environmental education in the schools of Shiga Prefecture, therefore, has the following features:
1) each school in every region is involved;
2) environmental education takes place from a world perspective; and
3) environmental education is based on the children's direct experience.

The Problems of Environmental Education in Schools

The main issue in environmental education in schools is to make children aware of the importance of environmental issues. It is necessary to improve the quality of this education to allow people to face these issues and to be able to deal with them concretely. This relates to the following four points:

Teaching methods and techniques

Though teachers practice environmental education in subject studies, the issue remains of how to actually solve environmental problems. The following points are related to raising awareness.
1) *Improvement and creation of teaching materials and tools.* Passive learning is not sufficient to promote an eagerness for study in children. It is important to improve their ability to understand environmental problems for themselves and to think about how to solve them. It is, therefore, necessary to improve teaching materials and tools.
2) *Nurturing of problem-solving ability and expressiveness.* Environmental education is commonly practiced through activities involving direct experience in school. Experiential learning is worthwhile not only to allow direct experience, but also to nurture the ability to assess and solve problems. It is necessary to improve this type of education.
3) *Enabling participatory learning.* There are many ways to understand and solve the present environmental problems, and so it is necessary to find more than one solution. It is important to recognize that different people have different values, and to learn to work with them. For these purposes, it is necessary to improve participatory learning through techniques such as role playing or debates.

4) *Education Providing Skills for the Solution of Environmental Problems.* At the present, teachers tend to teach scientific knowledge of the environment, but this knowledge is insufficient without practical skills. As an example, in order to solve environmental problems by decreasing the burden on the environment, it is necessary to decrease waste, and practical ways can be taught of achieving this.

I offer two suggestions for improving teaching methods.
1) *Improvement of teaching methods and techniques.* At present, instructors teach mainly subject matter, and tend to teach their attitudes on countermeasures. It is, therefore, necessary to focus the study on practical skills in the learning process.
2) *Teaching related subjects.* It takes time to study the practical skills using the children's own ideas. If teachers don't have enough time for that in subject lessons, they must increase the opportunities to study this in special curricular activities. Thus, in order to think more deeply and systematically about the environment, it is necessary to have a comprehensive interdisciplinary curriculum bringing together related subjects.

Improving educational conditions
It is necessary to improve educational conditions to promote environmental education. I mention three important problems here.
1) *Saturday holidays.* If teachers are to practice comprehensive environmental education, they have to use the regular school timetable. In the past this was easy, but now that there are two Saturday holidays a month now, the regular timetable is often devoted to other subjects. The distinctiveness of the timetable in each school is reduced. In the present situation, many schools integrate environmental education with other subjects, and the content is restricted by timetable constraints.
2) *Environmental education in integrated studies periods.* The first report by the Central Council for Education reports on the educational problems of Japan in the 21st Century, and states that environmental education classes must be taught in the integrated studies period, as other periods will be devoted to specific subjects. This period will be crucial for the development of environmental education in the near future. At the present, the Curriculum Council considers two hours a week as appropriate for integrated studies. It is necessary to have regular hours for environmental education in this prefecture, so it is;

therefore, necessary to organize the textbook to allow the most effective teaching of environmental education in the integrated studies period.
3) *Providing school gardens with wildlife.* Up until now, people have thought that it was necessary for the school environment to be clean, so there was no wildlife in school gardens It is important to provide the provide space in these gardens for different kinds of wildlife.

Cooperation between the home and society

It is important for everyone to work as hard as possible to solve environmental problems. People cannot receive their education only at school. It is important to promote education in cooperation with home and society, and it is necessary to determine the role which should be assigned to each within the context of life-long study.

Ethical problems of environmental education

Today's main problems concerning the younger generation are a result of their declining system of values. If the young people cannot develop a value system, the country will change dramatically in the future. However, there is no clear way to solve the problem because we do not have a common national standard of values that correspond to how we live and what we should be. Thus, our current values are in a state of disorder, and today's society is undergoing tremendous change. In order to solve this problem it is necessary to educate and develop the values of young people.

As human beings' activities cause global environmental problems, we have become aware that we must decrease the burden on the environment. We do not only live for ourselves. We inherit a legacy from our ancestors which we will pass on to future generations. This continuity can thus help us develop self-control and a system of environmental ethics.

I think that the main function of environmental education is to develop children's consciousness, and teachers must teach them that the earth's condition is critical, by improving the quality of environmental education. It is also necessary for the educational administration to support this process.

Selected Bibliography

Shiga Prefecture. 1974. *Handbook for Environmental Preservation.* Shiga Prefectural Government, Otsu, Japan (in Japanese).

Shiga Prefecture. 1976. *Collections of Good Activities for Environmental Education.* Shiga Prefectural Government, Otsu, Japan (in Japanese).

Shiga Prefecture. 1980. *Subtextbook for Environmental Education.* Shiga Prefectural Government, Otsu, Japan (in Japanese).

Chapter 10

The Development of Environmental Education Curricula and Some Examples of Thai Experiences

Sirmsree Chaisorn, Faculty of Education, Chiang Mai University, Chiang Mai, 50200, Thailand.

Introduction

In schools, students can receive environmental education from both course-related learning and co-curricular learning. A big gap can still be found between students' knowledge of environmental education and their actions to protect their surroundings. Environmental education should be further emphasized and strengthened.

Young people's awareness of environmental issues and problems should be enhanced throughout our increasingly interdependent world as a matter of urgency. Besides non-formal education which can take place anywhere, school curriculum development seems to be one of the most promising channels for addressing these concerns since school life begins very early and lasts a minimum of nine years for each individual in most countries. School learning is gained both through course work experience and in other ways. The latter include co-curricular activities and the arrangement of school physical and social environments. Many aspects of learning can be extended to a community-based inquiry or school involvement in helping to solve community problems.

Approaches to Developing Environmental Education in Schools

Integration through infusion

This approach calls for all subject teachers to integrate environmental facts and concepts into the courses they are teaching whenever it is possible and appropriate. Some topics in all courses lend themselves well to such modification or application. This is one way to revitalize the prevailing courses in the school syllabi so that they become more relevant to students' lives and communities. Therefore, updated information about local, regional, and world environmental situations must somehow be collected, simplified, and disseminated to teachers in all subject areas. Who should take on this responsibility? Governmental organizations, non-governmental organizations working to upgrade environmental quality or sustainable development, and especially educational organizations at all levels should give more attention to these activities. As teachers become more aware and better equipped with relevant environmental data, they will take more of the initiative in organizing learning experiences for their students in relation to their own subjects. Concerned teachers will use environmental content as the means for students to derive knowledge and practice specialized skills in their subject areas or disciplines. One problem is how to make every teacher realize the importance of integrating environmental concepts or data in their own disciplinary subjects. As some curriculum experts say, when everybody thinks it should be the responsibility of everyone else, then no one will take responsibility. It might turn out that only a few teachers use related environmental data or concepts to teach the principles of their own subjects. Moreover, many teachers may be concerned that repetition across subject areas may occur if everybody teaches about the environment. In this regard, one can envision many aspects of the environment which can appropriately be promoted or emphasized in accordance with each subject specialization. This curriculum development approach is mentioned when we discuss who should teach moral education or ethics.

Integration through specifically-designed integrated units or courses

This more popular approach may or may not need interdisciplinary teaching teams. Lately, many people concerned with curriculum development in different countries have tried to include "environmental studies" in school syllabi, usually in either science or social studies programs. The Institute for the Promotion of Science and Technology in Thailand, for instance, is responsible for school science curriculum development. They have introduced environmental studies as part of the required general science subjects, and as an elective course called "The environment surrounding us" in the lower-secondary school syllabus. Earlier this year, as the cry for environmental education became louder, a new required course dealing directly with environmental data and concepts was introduced for upper secondary school students. In the social science area, besides some environmental content in students' required courses, there are many elective courses with titles like "Our local community" in Thailand's syllabus for lower-secondary school students. Science and social studies teachers admit that there is some repetition, but they are persisting with specialized courses as the basis for environmental studies. Most environmental education specialists, however, call for a more integrative or holistic outlook. Such programs will more and more be placed in the hands of environmental educators whose concerns are not just science or social science. At present, when there are not enough environmental educators, interdisciplinary planning teams are recommended to handle these problem-based units or courses. Theoretically, the approach is sound and acceptable throughout the academic world. Practically, not many instructors have used this approach. Cooperation among people in different fields requires especially broad-minded, enthusiastic school personnel with initiative, especially at the secondary school level where teachers are departmentalized. Many research projects on environmental education are being conducted by master's students at Chiang Mai University. As they have concentrated on designing integrated units with interdisciplinary team teaching, both teachers and students have had some exciting experiences. When two or more teachers work cooperatively to plan and organize student learning, it tends to promote a more interesting and meaningful experience. Students can seek advice from more experienced people, and will themselves learn to work cooperatively. The team teaching approach is not necessarily interdisciplinary. Personnel within the same fields can plan team teaching as well. Again, this is not regularly practiced due to most people's independent work habits.

The research findings showed that team teachers felt that the existing schedule imposed time constraints, and they requested more flexible and supportive school policies. Whether teaching is in teams or not, units or courses designed for environmental education should provide a variety of teaching strategies, learning experiences, and independent learning opportunities for students, as well as fostering experimental and inquiring attitudes. School personnel should be more concerned with whether students can correctly answer questions about how to conserve our environment, or whether they can discuss the concepts of the various "Rs" for environmental quality, but they have still not put these ideas into action either at home or further afield.

Co-curricular activities

In Thailand, many schools have had successful experiences in organizing environmental education through special school clubs for nature study, environmental conservation, etc.. Students in many schools have enjoyed participating in "Recycle/Reuse" projects. They have devised new products made from different materials. Students in agriculture promotion clubs have prepared natural fertilizers for plants. Some clubs have worked on providing students with essential consumer knowledge. Displays or exhibitions of students' green products have become annual events in some schools. With support from school administrators and the initiatives of school personnel, their knowledge and attitudes should be widely extended to students throughout the school and not just a few interested groups.

The International Lake Environment Committee (ILEC) school environmental education projects carried out in a few elementary and secondary schools in the provinces of Chiang Mai and Pattani in Thailand have demonstrated the success of both coursework and co-curricular activities in terms of students' experiences and teachers' cooperation. Also, recently, other supportive units including regional education offices and some NGOs have organized many projects that enable students to be more attentive to the environment. For example, they have given prizes to schools with good environmental projects. Many schools have tried to create a good physical and natural environment for their students to live in. And in most schools, both teachers and students participate in these activities.

Environmental problems can be seen in all parts of the country because of inappropriate methods of garbage disposal, polluted water, polluted air, excessive use of energy, deforestation, etc. However, our efforts must continue. There is a great need for both vision and action to promote environmental consciousness, especially for those of us in education.

Selected Bibliography

ILEC. 1996. *Final Report on Promotion of Environmental Education in Developing Countries(1991-1995)*. International Lake Environment Committee Foundation, Kusatsu, Japan.

Jørgensen, S.E., M. Kawashima, and T. Kira, Eds. 1997. *A Focus on Lakes/Rivers in Environmental Education (Guidelines of Lake Management)*. International Lake Environment Committee Foundation, Kusatsu, Japan.

Ramsey, J.M., H.R. Hungerford, and T.L. Volk. 1992. Environmental education in the K-12 Curriculum: finding a niche. *Journal of Environmental Education*. 23(2): 35-45.

Talisayon, V.M. 1990. *Community-based Environmental Education. Sourcebook in Environmental Education for Secondary School Teachers*. UNESCO, Paris, France.

Wilson, R.A. 1993. Educators for earth: a guide for early childhood instruction. *Journal of Environmental Education.* 24 (2): 15-21.

Chapter 11

In-Service Training Programs in Environmental Education in the Philippines — The Role of UP-ISMED

Merle C. Tan, Institute for Science and Mathematics Education Development at the University of the Philippines, Diliman, Quezon City, 1101, Philippines.

Introduction

Environmental Education in the Philippines has been going on for many decades. Many government and non-government agencies have integrated environmental education concepts into their programs. Despite these initiatives, environmental degradation continued unabated. The 1991 report on the state of environmental education in the country (a study headed by the author and funded by the Asian Development Bank) summarized the reasons: the activities were not cohesive and lacked direction; only a small percentage of teachers were doing environmental education; only a few had undergone formal training in environmental education; and there was a dearth of instructional materials that focus on the development of problem solving and other high-order thinking skills in environmental education.

To address the concerns in the status report, the environmental education team formulated in 1992 a National Environmental Education Action Plan. The action plan dealt with strategies to improve the delivery of environmental education across sectors. One of the strategies recommended was the development of a curriculum framework to serve as a guide in the formulation of environmental education curriculum frameworks for basic, technical-vocational, and higher education, including teacher training and informal education.

It took five long years for the government to implement the recommendations in the action plan. In 1996, the environmental education team reviewed the status of environmental education in the country through field visits and the use of questionnaires. Based on the updated status report, a broad curriculum framework was formulated and then validated by environmental educators across sectors. A pilot training program for teachers was undertaken using the framework as the guide. The project also developed exemplar lessons and instructional aids which were tried out in schools to determine their suitability for different target audiences.

The Institute for Science and Mathematics Education Development at the University of the Philippines (UP-ISMED), the center for teacher training, curriculum development, and research in science and mathematics in the Philippines, plays an important role in promoting environmental consciousness both in formal education and elsewhere. Its environmental education-related training programs for teacher educators and classroom teachers include curriculum framework orientation, enhancement of skills in investigating or monitoring community environmental problems and issues, and utilization of these problems as the context for lessons. UP-ISMED has also developed TV lessons and radio plugs on environment topics. The philosophy is simple: when teachers have acquired the necessary knowledge, skills and attitudes toward the environment, they will be able to develop students' "hands-on, minds-on, and hearts-on skills" that will help them become responsible and caring citizens.

The Philippine Environment

Filipinos all over the country are witness to the deterioration of the Philippine environment. In one flash flood episode in the Central Philippines attributed to deforestation, about 5000 Filipinos died. On another occasion, tons of fish were found floating in Manila Bay. Solid wastes clog the waterways. Most surface waters especially near Metro Manila are silted and biologically dead. These environmental problems affect not only the health and productivity of the people but also the sustainability and integrity of our natural resources.

Caring for the environment and the life support system is everyone's responsibility. The Philippine government has launched a number of environment-related projects across various sectors throughout the whole country. The formal school system has been assigned a special role to develop science, technology, and environmentally literate citizens. Realizing that the success of an environmental education program is influenced greatly by the way the teacher implements it, the teachers have been challenged to play a major part as they interact with their students and act as role models in environmental protection, conservation, and improvement. All teachers are expected to become environmental educators.

How can teachers become effective environmental educators? What kind of training programs should teachers be exposed to so that they can develop environmentally-friendly citizens and high-order thinking individuals?

Environmental Education in the Philippines

Environmental Education in the Philippines gained official recognition with the promulgation of the Philippine Environmental Code in 1977, a step in support of the Belgrade Charter. The Code specified that the Department of Education, Culture and Sports (DECS) should integrate environment-related concepts into the formal school curricula at all educational levels. The Department of Environment and Natural Resources (DENR) was delegated to promote environmental education using formal/informal methods.

Despite this promulgation, environmental deterioration continued at a fast rate. Government and non-government programs seemed unsuccessful in alleviating the problems which were caused by the synergy of many phenomena, including the rapidly increasing population with its greater demands for a better quality of life, and the frequent natural calamities that visit the country. Many observed that the environment and economic development programs were in conflict with each other and did not address the basic problem of poverty. In addition, the programs were not cohesive and lacked direction (Asian Development Bank, 1991).

Based on the status report and the need to address the concerns articulated therein, a National Environmental Education Action Plan (NEEAP) was prepared in 1992. The major aim is to improve the delivery of environmental education across sectors to produce a mass base of environmentally aware citizens and develop/increase the scientific manpower to be involved in environmental management, protection and improvement.

The implementation of some of the priority projects described in the action plan started in 1996. The executing agencies are DECS and the DENR. One of the projects implemented was the development of a broad environmental education curriculum framework which now serves as a guide in developing environmental education programs nationally, across sectors. The broad curriculum framework was also used as a template for developing the environmental education curriculum frameworks for specific sectors: basic education, technical-vocational and higher education including teacher training, and informal education. It was also used as a basis for developing exemplar scope and sequence charts and lessons for the different target groups, as well as support materials like posters and videos for basic education. This framework and other relevant information are compiled in a one-volume curriculum guide.

The Environmental Education Curriculum Framework in Brief

The broad environmental education curriculum framework has four dimensions: objectives, content, learning experiences, and evaluation.

The objectives dimension

The students are expected to be able to do the following:
1) learn about the environment;
2) acquire skills to investigate the environment;
3) develop a feeling of concern for the environment; and
4) communicate the environment-related knowledge, skills, and attitudes through practice and/or modeling.

Specific indicators show that students have acquired these skills. The following list may not be complete but it gives ideas of the kinds of behaviors expected. For example, students have learned about the environment if they are able to:
1) identify its components;
2) classify the different levels;
3) analyze interrelationships within and between environments;
4) distinguish nonrenewable from renewable resources;
5) illustrate how technology has increased the potential impact of people on it; and
6) analyze the different actions that people can take to influence decisions on it.

Students have acquired the skills in investigating the environment if they are able to:
1) use all the senses to observe its components, and to measure and record data;
2) organize, classify, analyze, and evaluate data;
3) interview appropriate people on environmental topics to gain a variety of views;
4) interact with people in different communities and investigate their experiences in relation to the Philippine environment;
5) debate alternative viewpoints on environmental questions and issues;
6) design and implement balanced management plans for the environment; and
7) apply problem solving techniques to issues related to local and global environments.

Students have developed feelings for the environment if they are able to:
1) take responsibility for their behavior towards the environment;
2) respect the rights, needs and opinion of others;
3) demonstrate enthusiasm to investigate different aspects of the environment;
4) show an appreciation for the unique features of the Philippine environment; and
5) appreciate the knowledge and experiences of cultural minorities in managing the environment.

Students have acquired the skills of communicating their knowledge, skills, and attitudes if they practice the following behaviors:
1) reduce wastes, reuse and/or recycle materials;
2) participate in environmental protection campaigns;
3) help organize environment oriented organizations; and
4) express value judgment on environmental issues through lobbying, petitioning, and letter writing.

Different emphases are recommended for environmental education in various sectors and/or grade levels. The objective is to help young children acquire habits that are friendly to the environment. When they grow older, they are expected to become environmentally aware citizens. When they are more mature in age and experience and have acquired higher educational competencies, they are expected to exhibit behaviors appropriate for environmental professionals.

The content dimension

The environment covers the natural environment, the built environment, the social environment and the spatial environment (the last being concerned with space and time). Some environmental education curricula present the environmental components as unifying themes. Others use as themes elements of the life support system, such as air, water, land, energy, plants, and animals and their interactions. Still others focus on local, national, or global environmental problems.

There are some problems arising from the use of these themes, however. Overlapping of content is common as they are discussed in different subjects. The other concern is the superficial treatment of the topics in different learning areas.

The environmental education framework suggests the use of environmental principles and their core messages as unifying themes. The main reason is that these principles and core messages are applicable to any ecosystem, any environmental problem at any time or anywhere. In other words, these can be introduced in different contexts. They have to be internalized by learners as they go through different stages of maturity and education. They can be introduced spirally from simple to complex, from macro to micro depending on the target group, and the characteristics of the learners. The principles and the messages they convey are as follows:

1) *Interconnectedness / interdependence.* Everything is connected to everything else. What one does affects others, directly or indirectly.
2) *Diversity and stability.* All life forms and/or materials are important. The more diverse the system, the more stable it is.
3) *Change.* Everything changes. Some changes improve the environment; others degrade it. Changes that destroy the environment must be prevented.
4) *Balance of nature.* Nature knows best. It has its own laws to maintain itself.
5) *Finiteness of resources.* Most resources are nonrenewable. They must be used wisely and prudently.
6) *Population growth and carrying capacity* . A given ecosystem can only support a limited number of individuals, over a specific time. Beyond the carrying capacity level, an imbalance in the ecosystem may occur.
7) *Materials cycle.* Matter is neither created nor destroyed. It can only be transformed from one form to another. Matter goes somewhere. When it accumulates in one place, the material becomes a pollutant.
8) *Stewardship.* Human beings are part of nature. They are stewards of nature's gifts.
9) *Sustainable development.* Economic activities can be pursued but not at the expense of a degraded environment.

The learning experience dimension

This dimension describes the *how, where,* and *when* of environmental education. The learning experiences depend on the target group.

The *how* of environmental education involves learners in "hands-on, minds-on and hearts on" practical experiences. It emphasizes not only the development of cognitive and psychomotor skills but also affective skills. It promotes clarification of values and real experiences in problem solving and decision making.

The *where* of environmental education means bringing the learner on site, where the environmental problem or successful environmental education program exists. It also implies that environmental education topics be taught across subject areas. For the informal sector, environmental education has to be integrated into all relevant programs.

The *when* of environmental education is from early age to old age. Environmental education is an ongoing process; it is lifelong education. It could happen at home, in the school, in the workplace and in the community where the learner interacts.

The evaluation dimension

Evaluation is an integral component of learning. It is a means for modifying teaching to more closely meet the needs of the learners. All the dimensions of the curriculum, including the modes of integration, must be evaluated. The relationships between and among the dimensions of the environmental education curriculum framework are shown in Figure 1.

Teacher Training Programs in Environmental Education

Environmental educators must be familiar with the knowledge and pedagogical base of environmental education. Therefore, environmental education training programs must not only provide accurate and updated environmental data but should help teachers process information which is often conflicting to enable them to solve problems creatively and make intelligent decisions. Teachers who possess a holistic understanding of the interconnectedness between and among environmental systems can find solutions to problems faster than those who do not have this understanding. A high-order thinking teacher inspires students and the community to become one, too.

It is also important for teachers to understand the characteristics of learners at different grades/year levels so that they can adapt their teaching style and coverage of content to the target group. They should learn to develop environment related lessons relevant to the students experiences and know how, where and when to integrate these in the existing curricula.

Investments toward promoting environmental protection and improvement skills are visible in the number of environmental education activities in schools, in communities, on TV, radio and

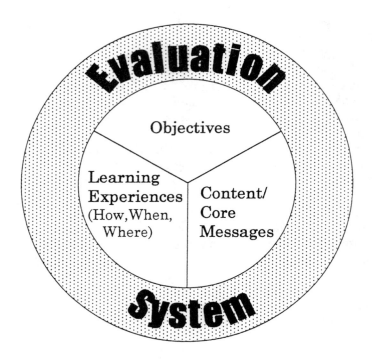

Figure 1. The environmental education curriculum framework dimensions.

other mass media going on throughout the year. On the other hand, the training of teachers both at the pre-service and in-service levels is continuously being organized by universities/colleges with support from professional teacher organizations and most often funded by industries and the private sector.

The Role of UP-ISMED

The Institute for Science and Mathematics Education Development is known for its innovative programs in science and mathematics. Established in 1964 as a Science Teaching Center with six science education specialists and a few administrative staff, it now has 45 academic staff in different fields of science specialization, and 60 support personnel. It has developed three generations of textbooks for students at the basic education level (1973, 1980, and 1989 respectively) and resource materials for teachers, both print and non-print. Its state-of-the-art facilities and equipment enable the staff to regularly organize and conduct international and national training programs for teachers and school administrators. It has been the leading agency in many international research projects in science and mathematics participated in by the Philippines. Its locally based research is being used in policy reforms in science and mathematics education. The Institute has won a number of awards for its community extension programs and other related projects. In most of these activities, the promotion of environmental education is a major component.

One of the current projects being implemented by UP-ISMED is the Science and Mathematics Education Manpower Development Project (SMEMDP), funded by the Government of Japan. The SMEMDP is focused on the enhancement of practical work and high-order thinking skills of UP-ISMED staff so that they, in turn, can effectively transfer the skills to science and mathematics teachers and educators across the country.

There are seven subject areas involved in the SMEMDP, namely Elementary School Science, Elementary School Mathematics, High School Earth Science, High School Biology, High School Chemistry, High School Physics and High School Mathematics. Environmental Education integration is not limited to Earth Science alone. All the subject areas incorporate environment-related concepts and skills in the design of practical work and concept formation activities.

Currently, the Institute is implementing a community-based project entitled "River Watch." One component involves monitoring the physical, chemical and biological quality of the river near the University. The other component aims to determine the effect of the environment-oriented instructional materials and the project itself on the knowledge, skills and attitudes of students in schools near the river. The project is being implemented through the cooperative efforts of the Local Government Unit, the DECS and other relevant Institutes in the University.

UP-ISMED opened a resource learning facility in November, 1997. The area for teachers includes an environmental diorama which dramatizes the interconnectedness between the forest and the sea. Exemplar lessons that demonstrate how the diorama can be used for teaching can be availed of by visitors. An attachment program is being studied to enable selected teachers to stay at the facility and develop instructional materials for classroom use. Staff of the Institute will serve as advisers. The area for students contains interactive activities to stimulate students' interest in science, mathematics, environmental studies and technology-related careers.

Figure 2. The environmental education curriculum tree.
HOTS: High Order Thinking Skill.

Conclusion

The environmental problems together with the social, political and technological changes going on in the Philippines have created greater demands and challenges for the education sector. The school curriculum has always been expected to reflect these changes. Accordingly, the different levels of education in the country have modified their curricula in order to meet these new demands. One of the strategies used to respond to these demands for change is the integration of societal issues including environmental issues into the curriculum. Teacher training programs have been designed to help teachers integrate these issues into their respective learning areas and become effective environmental educators.

Environmental education is a dynamic force to halt environmental degradation and depletion of the country's natural resources. But the environmental education activities have to be cohesive and integrated.

This paper describes the environmental education initiatives in the Philippines. It also gives emphasis to the environmental education curriculum framework. Figure 2 illustrates the important points.

Put simply, the tree represents a holistic environmental education program. The curriculum (the roots) provides the anchor and stability for the program. The teacher (trunk) connects the curriculum and the target groups. Environmental education must be integrated into all levels of the educational ladder (the branches) — preschool, basic education, higher education and beyond. The teaching strategies (the leaves) must be varied and interactive to develop high-order thinking students (the fruits). The tree must be provided with enough sunlight and water. In educational terms, these elements of the support system must come from school administrators, government and non-government agencies, and other sectors of the community.

References

Asian Development Bank. 1991. *The State of Environmental Education in the Philippines. A Report of the Project on Environmental Education in the Philippines. Phase 1.* Asian Development Bank, Manila, The Philippines.

Asian Development Bank. 1992. *The National Environmental Education Action Plan. A Report of the Project on Environmental Education in the Philippines. Phase 2.* Asian Development Bank, Manila, The Philippine.

Chapter 12

Local Involvement through the Environmental Network: The Key to Successful Natural Conservation

Suraphol Sudara, Department of Marine Science, Faculty of Science, Chulalongkorn University, Bangkok 10330, Thailand.

Introduction

Thailand, a country very rich in natural resources, in the past due to her unique geographical location now faces problems of environmental degradation and natural resource over-exploitation. During the past two decades, the country's forest coverage which used to be over 60%, has been reduced to less than 16%. With such a reduction, droughts and floods have become very common.

Rich agricultural lands, particularly in the central plain, are being degraded due to new land uses which introduce pollution contamination as well as some modern agricultural practices such as monoculture farming, mismanaged water irrigation, etc. These have led to a degradation in soil quality and an explosion in the output of agricultural pests.

The Gulf of Thailand used to be one of the richest fisheries in the world. Due to improper management, the number of bottom trawlers and push-net boats with powerful engines increased tremendously, leading to over fishing. Trawlers and push-net boats came within the 3 km limit. This intrusion together with the tremendous expansion of shrimp ponds has caused the destruction of coastal ecosystems, particularly seagrass beds and mangroves. The fisherfolk who depend on these coastal ecosystems are now faced with hardship.

Informal Education

Education is a very important tool in the fight against environmental degradation. At present the Thai education system recognizes the importance of these problems and subjects concerning the environment are now offered at all levels of education from primary school through university levels.

Informal education is very important to create awareness of environmental issues. The use of mass media, both printed and electronic, is essential. Environmental awareness is very important as people are the primary source of the problem. If they are not actively involved in cleaning it up, at least they should not cause any additional problems. Awareness generated through the mass media still may not result in action when it is needed. One difficulty is in convincing the mass media sector to propagate the message of environmental awareness.

The Activities of Buddhist Monks

Buddhism is the main religion in Thailand, and the monks usually play a very important role in the local communities. Environmental education has been assisted by the efforts of the NGOs, such as the Siam Environment Club, the Wildlife Fund of Thailand, and the Religion for Society Foundation, which realizes that the teaching of the Buddha rests mainly on an understanding and respect for nature. There have been many contacts with some of the well known monks who have been preaching about this matter for many years, and lots of books have also been produced as a result of these efforts.

Many active young monks have been using their regular meetings to provide informal education on the protection of natural resources. The NGOs also help by organizing special seminars on environmental awareness for monks. When the monks return to their areas, their preaching encourages villagers to follow and start practicing environmental protection. This is a very powerful type of informal education. Many monks inspire villagers to protect the nearby forests or to set up protection measures for their water resources.

Some monks use the tactic of wrapping saffron cloth around big trees imitating the ceremony in which a young man becomes a monk. This activity symbolizes that all of those trees are already under special protection which others should respect. This activity has been very successful on many occasions. The latest was a special event to celebrate the 50th anniversary of the accession to the throne of His Majesty the King. About million trees within a very large area in the north of Thailand have been declared sacred and will not be cut down.

The Projects of Her Majesty the Queen

Her Majesty the Queen, has been campaigning very actively to protect forests. She has started many projects such as the Forest Love Water Project, the Small House in the Dense Forest Project, etc. She has emphasized the important relation between forest and water as the forest functions as the main source of the water supply. Some of the rehabilitation projects have been very successful.

Community Protected Forests

Many NGOs now focus their efforts on the development of community based activities for forest protection. They believe that the most effective means of reforestation is to accept and support local community rights and their ability to conserve the forest, particularly where communities already support forest conservation.

In 1989, a dispute arose in the northern province of Chiang Mai between villagers and a politician. Then villagers consider the disputed land, approximately 200 ha, to be a communal forest as it is still in a natural condition even though it is not a virgin forest. The politician claims that it is a degraded forest, and he would like to give a concession to build a resort there. This conflict caught national attention through media reporting. Finally, the Royal Forestry Department settled the matter by awarding the villagers the right to manage the forest. However, there are no legal channels for enabling local communities to manage a forest.

It is clear that community protected forests are being established when a resource such as land, forest or water that is vital to the community is being threatened. In northern Thailand the "muang fai" system of irrigation and protected communal watershed forests was established by adapting an old tradition in response to threats of deforestation which would decrease the availability of water, which is very important to rice cultivation. However, communities usually do not conserve the forest until the level of resource decline actually threatens their survival.

In Jomtong District, Chiang Mai, villagers who live in the valley noticed that their natural water supply had been declining. They went up the mountain and found that the forest on top had been destroyed by a hill tribe practicing slash-and-burn cultivation. The villagers had to physically fence off and patrol the forest in order to protect their water supply. At first, the local authorities misunderstood and thought they were causing unrest. It took some time to clear up the matter. Finally, the forest could be protected.

Also in Chiang Mai, Wat Chan is a well known case where the villagers had been protecting the natural pine forest which they normally depended on for natural products. Foreign experts in forest management had advised the Royal Forestry Department to replace the mixed species of the natural pine forest with one commercial pine species. Villagers and NGOs opposed the plan and asked that their natural pine forest be kept in its natural condition. Finally, the villagers won and they still have their natural environment.

A local NGO in Nan Province, Huk Muang Nan, had long been educating villagers to realize the importance of the natural forest to their livelihood. After a few years, the villagers were successful in protecting and rehabilitating the Nan forest. This proved that the hard work in public education was having a positive effect.

In many areas, people selected deteriorating forests for their new settlements. Rather than clearing the land for agriculture as they used to do, they now follow a new approach, agro-forestry. They cultivate their crops among the big trees to avoid deforestation. This practice is now widely accepted. Even though the forests have not recovered to their former condition, they perform a more useful ecological function than if they had been cleared.

One village headman, Viboon Kemchalerm, of Chacherngsao Province, has successfully demonstrated that his natural integrated species approach to agriculture could very well improve the village economy. His approach is to cultivate different varieties of vegetation and trees according to the natural conditions. As all of those species compete with each other, the result is an improvement in the ecosystem. Products can be collected regularly because the system is already sustainable. In addition to the trees and vegetables which are used regularly, medicinal herbs are also planted. This form of sustainable agriculture depends on the natural selection process, and it is now spreading to many areas.

Coastal Conservation and Local Involvement

Since 1987, the need for conservation activities with villagers' participation has been widely recognized. In the hope of spreading ecological awareness among the coastal communities the Yad Fon Foundation began replanting mangroves in Trang Province. After that seagrass conservation practices were introduced to many coastal villages located near seagrass beds. Seagrass conservation protects seagrass beds from being destroyed by push-nets or trawlers. Eventually, they found that their catches improved and that fishing was easier. It made them realize the importance of their conservation efforts. They also sought help from the authorities to protect the seagrass and expand their area of protection. Other villages followed.

The record of catches in waters around the seagrass beds shows that recovery from destruction of the beds could occur within less than one year and that the catches are higher after the destruction stops. A large push-net can destroy about 80 kg of seagrass per hour, and around 84% of the catch would be composed of juveniles of economically important fishes, squids, crabs, and shrimps, all of which would be considered trash. Stopping the destruction of the seagrass beds could help conserve the rich variety of marine products.

The work of NGOs at Trang also helped to bridge the gap between local people and local authorities. Starting with a mangrove rehabilitation program, their efforts expanded to seagrass. Scientists' introduction of seagrass conservation practices to the coastal villages proved to be very successful after the villagers realized that their catches improved. On the insistence of the NGOs that they should preserve their fishing territories the villagers decided to extend their sphere of conservation to include the coral reefs. They set up buoys to mark the boundaries of coral reefs in their area. After a while, conservation became a regular part of the villagers' activities. They extended their efforts to conservation programs for marine turtles and marine mammals such as dugongs, sitting whales, and dolphins (Sudara et al., 1994).

The Holistic Approach in Conservation

Fortes (1994) suggested that material export, high inter-habitat similarities in fish, crustaceans, and epiphytic alga community composition provide strong evidence to suggest close functional linkages between mangrove and seagrass ecosystems. Poovachiranon and Satapoomin (1994), reported in Thailand that of 71 species of fish captured from a seagrass bed and 69 species found in the adjacent mangrove creek, 45 of these species were found in both habitats.

From the reports of fishermen in Petchburi and Samut Songkarm Provinces in Thailand, it appears that the cockle culture on the mudflat in front of the mangroves declined after the mangroves had mostly been cleared for shrimp pond construction. Since the conservation activities of the local people started a few years ago, and after the successful rehabilitation of the mangroves, the population of cockles has been increasing. The cockle seed which used to be so poor that they had to buy seed from other areas, has now become richer, and they do not have to buy it any more. Nutrients from the mangroves help produce enough food for cockles to grow in larger quantity than before in these areas.

Sudara et al. (1992) demonstrated the interrelationship of fish communities between coral reefs and seagrass beds. The movement of reef fishes to seagrass beds demonstrates the connection between these habitats.

It is very well known that many fauna of the mangrove or the reef, i.e. mud crabs, groupers, etc., migrate out to sea to release their eggs and larvae in the plankton states. Those mesoplankton, the larvae, are transported widely. When they reach the juvenile stage, they usually remain in the seagrass beds which provide them with better shelter from predators as well as a better food supply. When they approach the adult stage, they either migrate to the mangrove or to the coral reef as their final habitat. If conservation activities in each of these habitats were dealt with separately, the ontogenic cycle of these fauna could not be completed because they need all those habitats. Usually only the cycle within each habitat is considered, but many species require different habitats at different stages of development. Hence, in successful conservation the activities in all these habitats should be treated holistically because all are interrelated.

In some cases, where all three ecosystems do not exist in a coastal area, the life cycle of those species may be different. They utilize only what is available, and the success of the community in such areas, as well as the community composition, will be different.

The Co-management Approach

NGOs have also played a very important role in getting authorities, such as the Office of the National Environment Board (ONEB) and the Fisheries Department to assist villagers in protecting their areas from the intrusion of trawlers and large push-net boats. Legally, these boats cannot operate within 3 km of the shore. This example illustrates the cooperation between the local villagers and the authorities to strengthen law enforcement.

The success story at Trang spread among the coastal villages. Starting with five villages at Trang, similar conservation activities spread to more than 50 villages in the south by around 1990. Yad Fon and Wildlife Fund of Thailand played the leading role in organizing meetings of villagers, scientists, and government authorities in order to strengthen local conservation practices. Their efforts resulted in the establishment of the Southern Artisanal Fisherman Federation, which provides an opportunity for local fishermen to get together to exchange information and voice their demands. Through the federation, they have asked the authorities for more stringent enforcement of the 3 km limit and for various kinds of assistance, such as to campaign on their behalf against certain policies and practices. The fishermen are also free to discuss their various needs with the authorities. In some areas they want the Fisheries Department to construct artificial reefs which would not only attract fish but also obstruct the operation of trawlers and push-net boats. In other areas, however, they have voiced their concern that artificial reefs would make it more difficult for them to fish by traditional methods. Clearly, there is a growing sense of local participation in coastal planning.

In 1995, the local fishermen, the governors, and the Fisheries Departments of three provinces, Phuket, Phangna and Krabi, signed agreements with the Deputy Minister of Agriculture to develop joint efforts in conserving natural resources within their respective areas. The villagers even built a center for officers to patrol the area.

Still another influential sector in coastal zone management is the private investors who are bringing increasing economic pressure. Industrialization, tourism, land development, agriculture, and even aquaculture, can all produce tremendous changes in land use patterns, cause degradation to the coastal areas, and change the local lifestyle. It is, therefore, very important that the investment sector be involved in coastal zone management to avoid conflict with environmentalist and conservationist groups. NGOs are bridging the gap by involving private investors in the process of co-management. They are invited to participate in discussions to solve conflicts or to plan management strategies.

It is hoped that all five sectors will be involved: local people, government authorities, scientists or academicians, private investors, and, most importantly, NGOs who serve as the catalysts, facilitators and initiators. If they unite their efforts and meet to discuss their needs and sources of conflict, they may be able to come up with workable solutions for a coastal zone management plan.

References

Fortes, M.D. 1994. Philippines seagrasses: status and perspectives, in *Proceedings, Third ASEAN-Australia Symposium on Living Coastal Resources, Vol. 1: Status Review*, Wilkinson, C.R., Sudara, S. and Chou, L. M., Eds., Chulalongkorn University, Bangkok, Thailand, pp 291 - 310.

Poovachiranon, S. and U. Satapoomin. 1994. Occurrence of fish fauna associated in mangrove-seagrass habitats during the wet season, Phuket, Thailand, in *Proceedings, Third ASEAN-Australia Symposium on Living Coastal Resources, Vol. 2: Research Papers*, Sudara, S., Wilkinson, C.R., and Chou, L. M., Eds., Chulalongkorn University, Bangkok, Thailand, pp 539-550..

Sudara, S., S. Satumanatpan, and S. Nateekanjanalarp. 1992. A Study of The Interrelationships of Fish Communities Between Coral

Reefs and Seagrass Beds, in *Proceedings, Third ASEAN- Science and Technology Week, Vol. 6: Marine Science: Living Coastal Resources,* Chou, L. M. and Wilkinson, C.R., Eds., Singapore, pp 321 - 326.

Sudara, S., S. Satumanatpan, and S. Nateekanjanalarp, 1994. Please Help Protect Dugong. *Thailand Illustrated.* 12(34): 32-36.

Chapter 13

Environmental Education in Support of Sustainable Development: The Case of Indonesia

Mohamad Soerjani, National Research Council and Institute for Environmental Education and Development, Puri Building, Jalan Warung Jati Barat 75 A, Jakarta Selatan 12740, Indonesia.

Environmental Degradation

There is a global trend of industrialization, which is generally aimed at increasing the efficiency and effectiveness of production processes to produce more goods and services. This creates a tendency for technology to develop without consideration for human beings, promoting the use of machines and robotization, while neglecting the question of human dignity. In a country such as Indonesia, this may be in conflict with the capacities and cultures of the people. Apart from differences in their cultures, people also have significant differences in knowledge, skills, and educational background.

It is understood that to improve the welfare of the people, economic development is the key factor, supported by a relatively high growth rate in industrial development. Unfortunately, this economic orientation of industrial development is known to be the main cause of the environmental crisis through the application of *polluting technology* which, in turn stimulates *excessive consumption* (see also Paul Shaw in ADB, 1991: 11). These two phenomena produce considerable waste, and this causes the degradation of the quality of our environment.

Other factors that may affect the quality of the environment are: population increase, poverty, social unrest and, last but not least, the distorting effects of policies. Ultimately, the victims of the resulting crisis are human beings. There is a tendency towards the spread of human genetic mutations caused by the pollution from toxic wastes, and the contamination of foods with hazardous wastes, which lessens the resilience of the human race. Other known environmental degradation processes are the greenhouse effect, and the holes in the ozone layer that allow ultra-violet radiation to affect the health and survival of life on Earth, including humans. Increases in acid rain and air and water pollution have become common trends in cities and industrial estates. As Indonesia is a country rich in natural resources, the exploitation of forest resources and minerals is causing many problems due to the loss of rare species of plants and animals and the drastic changes in the ecosystem.

Another aspect of the degrading environmental quality is the way in which development may also negatively affect the economic and social environment. Because of gaps in technological knowledge and skills, most of the natural resource exploitation results from the trade in raw materials, such as copper and gold ores, timber, etc. This has to be corrected through the application of science and technology to promote value added processes that will increase the Gross Domestic Products (GDP) from the value added to goods rather from the value of the raw materials. The social problems raised by development are due to gaps in the participation of various groups in the community in the development process. Community participation in the planning, implementation, and evaluation of development projects is formally required by laws and regulations. However, the reality is different. There are gaps due to differences in educational opportunities, knowledge, and skills among the various groups in the community. Consequently, the optimal participation of a greater number of people is not possible, and this means that people cannot participate, which affects social stability and security. It has to be admitted that efforts to promote sustainable development in Indonesia to resolve these problems are being gradually implemented although the result is not yet as successful as was expected.

Sustainable Development

The concept of sustainable development was introduced in 1972 at the UN Conference on the Human Environment in Stockholm. The theme of this world gathering is reflected in the conference declaration, which said that man is both the creator and molder of the environment, and that the human environment is a major issue which affects the well-being of peoples and economic development throughout the world. Implicitly, development requires serious environmental consideration. The concept of sustainable development was taken up and further clarified by the meeting on World Conservation Strategy (IUCN) in 1980 (Bojo et al., 1990: 13).

It was taken up once more by the World Commission on Environment and Development (WCED, also known as the Brundtland Commission) in 1987, which defined it as "development which provides for the needs of the present generation without compromising the needs and aspirations of future generations."

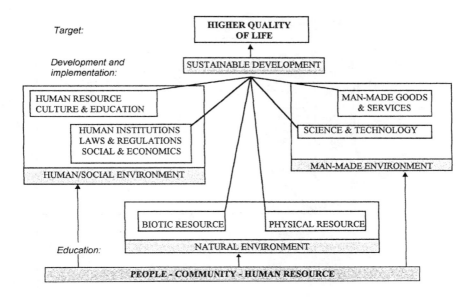

Figure 1. Sustainable development is to improve the quality of life through the wise management and utilization of development resources: the natural environment, the human or social environment and the man-made environment (modified from Soerjani, 1997a)

The quality of life

Following the Indonesian guidelines for national development, development is supposed to improve the quality of life. Improvement of the quality of life is defined as: an increase in life expectancy, which means a healthier community; alleviation of poverty, which means making people self-sufficient by providing for their basic needs; improvement of people's knowledge and skill through education; increase in people's participation in development activities, which means adequate employment opportunities; and proportional benefits from development results for everyone. See Figure 1.

Added value

A country or a system is sustainable if the depreciation of its natural capital in GDP percentage is more than the savings or

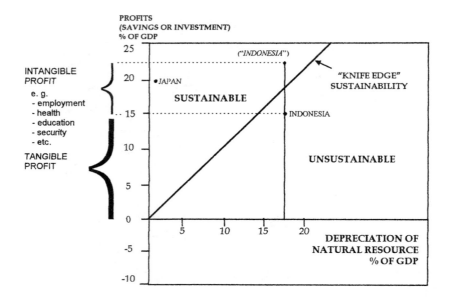

Figure 2. Sustainability of a country (according to Pearce and Atkinson, 1993) is based on the ratio between the netsavings and the depreciation of natural capital, as expressed in percentage of GDP.

investment. Indonesia is considered to be an unsustainable country because the rate of depreciation of its natural capital (17% of GDP) is more than its rate of net savings or investment (15% of GDP). This has to be overcome by the application of science and technology to process the natural resources with added values. There must be efforts to minimize the export of raw materials such as timber and other agricultural or forest products, mineral ores, coal, etc. The role of local manufacturing industries thus becomes that of processing them in order to add to their value.

Figure 2 poses a question which has to be answered. A country such as Japan appears highly sustainable, with a rate of savings of approximately 19% of GDP, and a rate of depreciation of natural capital of only 2%. The question is: how much in terms of raw materials is Japan importing from other countries including from Indonesia for its manufacturing industries? Another question is whether or not we can live in a peaceful and resilient world if its natural capital is being continuously depreciated. If a country is importing raw materials from an unsustainable country, where should the "knife edge of sustainability" in Figure 2 be located. In other words, how should the responsibility of conserving the world's natural resources be shared between the exporters and the importers of raw materials?

A further question is that of how this phenomenon of unequal partnerships between countries can be dealt with in environmental education for sustainable development.

Poverty alleviation and employment

Poverty is a global phenomenon. Victims exist both in both developing and industrialized countries. It is generally known that poor people are those who live in an intolerable habitat, deprived of adequate food, basic education, and health care, and who are mostly unemployed. Agenda 21 of the Rio UN Conference on Environment and Development in 1992 (Keating, 1993: 4-5) and Indonesian Agenda 21 (Anon, 1997: 12-13) show the same recognition that the struggle against poverty is the shared responsibility of all countries. It is recognized, however, that poverty has so many causes that no one single solution will solve all of the problems in every country.

There would be no more poverty if every individual or family was able to provide for their basic needs of food, health, and basic education. This means that people must work, be employed or create employment, which consequently requires equal opportunities for everyone to obtain appropriate education, enabling them to play a proportional role in sustainable development. Even though in reality all members of a community do not have the same abilities, they must be empowered to improve these.

Indonesia, is known as a country rich in natural resources, but it is also known to have more than 10% of its people living below the poverty line. To overcome this situation, there must be a gradual change in industrialization strategy. The emphasis on exports as a means of high economic growth must be balanced with a drive to enable Indonesia's own people to consume more of the products of the country's industry. In addition to increasing job opportunities and higher incomes, greater consumption of local products is an important indication of the alleviation of poverty (see Figure 3). With the increased control of the domestic economy, the external constraints of imports and exports can also be controlled, making the economy more resilient.

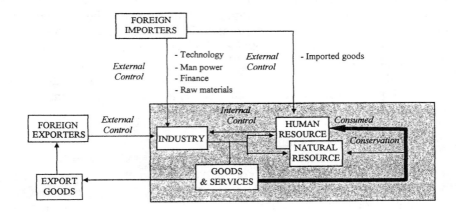

Figure 3. The maturity of a system or a country depends on how industry is under its internal or external control system (modified from Soerjani, 1997a: 21)

Education for Sustainable Development

Education plays a decisive role in human resource development. With the increase of the human population, the advancement of science, technology and arts, and the increase in the complexity of our environment, development will only be successful in improving the quality of life if it is implemented in a sustainable way.

The improvement of human resources through education is much more than an instrument for development. It is the ultimate objective of the development process. The advancement of science, technology, and arts in economic development creates, in many cases, a crisis of human redundancy, whereby continued economic growth apparently does not require the full participation of the available human resources. Education oriented towards sustainable development is able to put a human face on development to improve the quality of life(see UN-ESCAP, 1988: 7).

Education must be reoriented toward the new paradigm of sustainable development, and this must be integrated with the existing education system, from the pre-school level to the highest level of

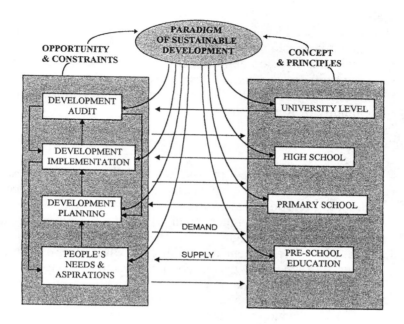

Figure 4. The paradigm of sustainable development must penetrate into various levels of education and various disciplines as well as penetrate into the various phases of development; the supply will be based on the demands (modified from Soerjani, 1993; Soerjani, 1997a: 21; 1997b: 46).

education (see Figure 4). The environmental principles in education itself should be based on the application of the concept of ecology (Soerjani, 1993; 145-160).

As shown in Figure 4 the paradigm of sustainable development is defined in terms of educational and academic concepts and principles, in addition to opportunities and constraints.

The paradigm of sustainable development will penetrate into all levels and all disciplines from pre-school education to the universities, and from the basic as well as the applied scientific disciplines, based on experiences in the implementation of development. On the other hand the concept and principles of sustainable development will provide guidance for its implementation and monitoring The result of development implementation must be reflected properly in the people's needs and aspirations.

Conclusions
1) Environmental education is an integral part of education for sustainable development.
2) Sustainable development must be implemented based on the decisive role of the human resources. People should, in turn, be the beneficiaries of development through the improvement of the quality of life (Figure 1).
3) Successes in sustainable development depend on the proper support of the social, natural, and man-made environments (Figure 1).
4) Sustainability is based on the efficiency of natural resource utilization, through the appropriate application of science and technology to add value to the resources exploited. The rate of savings should be higher than that of consumption, as a percentage of GDP (Figure 2).
5) Sustainable development must aim at strengthening the internal resilience of a country or its economic system, through a proper balance between the human perspective and the environmental perspective which takes in all life on Earth (Figure 3).
6) Sustainable development has to penetrate properly into all levels and all disciplines and provide guidance in the implementation of development based on the needs and aspirations of the people (Figure 4).

References
ADB. 1991. *Population Pressure and Natural Resource Management.* ADB Environment Paper No. 6. Asian Development Bank, Manila, The Philippines.

Anon. 1997. *Agenda 21 Indonesia National Strategy for Sustainable Development.* Office of the Minister of State for Environment and UNDP, Jakarta, Indonesia (in Indonesian).

Bojo, J., K. Miller and L. Unemo. 1990. *Environment and Development: An Economic Approach.* Kluwer Academic Publishers, Dordrecht, The Netherlands.

Keating, M. 1993. *Agenda for Change. A Plain Language Version of Agenda 21 and the Rio Agreements.* Center for Our Common Future, Geneva, Switzerland.

Pearce, D. and G. Atkinson. 1993. A measure of sustainable development. *Ecodecision*, pp 65.

Soerjani, M. 1993 Ecological concepts as a basis to environmental education in Indonesia. *Ecology in Education*, Hale, M. Ed., Cambridge University Press, Cambridge, UK, pp 145-60.

Soerjani, M. 1997a. Sustainable industrial technology with environmental outlook. *Science, Problems and Environmental Management*, Soerjani, M., Ed., Institute for Environmental Education and Development, Jakarta, Indonesia, pp. 25 (in Indonesian).

Soerjani, M. 1997b. *Development and Environment. The Idea and Implementation of Sustainable Development.* Institute for Environmental Education and Development, Jakarta, Indonesia, (in Indonesian).

Soerjani, M. 1997c. Degree programs in environmental science. *Environmental Education for Biodiversity and Sustainable Development*, Soerjani, M. and Hale, M. Eds., University of Indonesia, Jakarta, Indonesia, pp 97-119.

UNCED. 1992. *The Global Partnership for Environment and Development.* UNCED Secretariat, Geneva, Switzerland.

UN-ESCAP. 1988. Jakarta Plan of Action on Human Resources Development in the ESCAP Region. United Nations Economic and Social Commission for Asia and the Pacific, Bangkok, Thailand.

Chapter 14

Literacy Regarding Environmental Issues — Allonomy or Autonomy?

Tomitaro Sueishi, School of Environmental Studies, The University of Shiga Prefecture, 2500 Hassaka-cho, Hikone, Shiga Prefecture 522-8533, Japan.

Introduction

To begin with, I would like to describe the shortcomings of the present fashion in environmental education, mainly supported by the traditional sciences and the conventional usage of computer technology.

What are the objectives of environmental education? Do they include:
1) teaching people about the environmental predicament;
2) instructing them scientifically about environmental cause and effect relationships;
3) guiding them in adopting an "earth-friendly" lifestyle, including resource recycling;
4) allowing them to achieve a balance between ecology and economy with the familiar message of "think globally and act locally;" or
5) asking the help of professors of environmental ethics?

My answers to the above questions are partly "yes," but I must say "no" when I observe the speech and behavior of international policy makers who make wide use of information similar to the above. Too many international events have been held in recent years that were aimed at avoiding the coming global crisis, but it is well known that the gap between the northern and southern countries in both economic and environmental conditions has been enlarged since UNCED in 1992.

In December, 1997, COP3 opened in Kyoto, and the Japanese government was expected to take a lead in the conference. However, newspapers reported that Japan could neither propose numerical standards for CO_2 emission reduction nor lead the international community into a new phase of environmental sustainability.

If it is assumed that university professors have a duty to be involved in environmental education, it is often thought that if they simply join an environmental faculty armed with the conventional wisdom, then this will be adequate to bridge the gap between the campus and the real world outside. Otherwise, they retain a monopoly control over esoteric knowledge, expensive research facilities, and systems for the collection of sensitive information to which students have no access. So the first point for discussion in the next section is the difference between problems which have to be solved by specialists, and issues which must be solved through the mutual consent or participation in decision-making of all stakeholders.

Very closely correlated to this first point is a second point, that of judging when the time is right to move from the analysis of environmental phenomena and problems to decision-making or planning. This is dealt with in the third section which presents some case studies in Shiga Prefecture, and we must pay attention to the differences between programming, design, and planning. If the objectives of environmental education are defined so as to create better environmental conditions along with a desired economic growth rate, it will only require a repertoire of particular programming tactics.

In Japan, however, the programming procedures mainly imported from the United States were not valid due mainly to the local political situation. Regional designers who were believed to have a creative talent were often appointed to sketch the future plans for a region. In my opinion, international governments are in a similarly embarrassing maze of decision-making, especially with regard to environmental issues, which have typically resulted in a compromise solution called "sustainable development."

Contained in the discussion above is the very important problem of the language used to represent complex environmental connections and to discuss long-range cost/benefit relationships. The rapid progress in the performance of information systems at present would seem to provide a solution to this problem, but I think that we are not using the systems resourcefully and are not studying sufficiently the question of what communication really is. I will deal with this subject in the section on environmental informatics below.

Taking all these points together, I venture to propose a new concept of education, as the 3Ls+1M (literacy, linkage and leadership + mediation). The old fashioned view of education was restricted only to literacy in its narrow sense, the so-called "3Rs" (reading, writing and reckoning). This will be main subject of the later section of the paper on environmental communication.

What is communication? How is it different from conversation? Decision-making or the formation of consensus, coupled with the autonomous development of networks of citizens, will really be the central feature of the post-modern university. Universities at present seem only to be interested in recruiting lifetime students for environmental education. At the same time they fail to inspire youngsters sufficiently to go to the countries of the "south" to engage in activities which will educate both themselves and future generations.

The Difference between Problems and Issues of Concern for the Environment

In this section I draw on my earlier report on the same subject (Sueishi, 1990).

As the word "problem" suggests an undesirable condition compared with a desirable one, there are many environmental problems surrounding us, such as pollution, acid rain, the greenhouse effect, the thinning of the ozone layer, desertification, the destruction of rain forests, border incidents caused by hazardous wastes, and population growth which is too rapid for the environmental capacity (Sueishi, 1975). When these problems are identified, research specialists or statesmen are generally appointed to take part in solving them.

However, it is ordinarily observed that the addresses or lectures made by the people responsible only emphasize the seriousness of the problems, listing the various causes inherent in modern society and adding the problems of "how to solve them." In the case of research specialists, they can, of course, predict quantitatively the variations in the symptoms of the problem based on selected scenarios related to their origins. Nowadays attention is focused on the future rise in sea level which corresponds to CO_2 emission control. Despite their diligence, it is impossible to decide on a single scenario, because there are too many uncertain and unpredictable factors which they have probably not taken into account in a simple model.

Having responsibility for obtaining a perfect solution to the problem is like being accused of a criminal offense. When the specialist responsible fails and is blamed unfavorably, he will defend himself by saying that the result was inevitable and was caused by factors outside his control (Shimizu, 1972). People who try to solve problems ultimately have neither freedom nor responsibility: they are forced to deal with the same problems over and over again, especially complex environmental problems. This is a world of science which is unconcerned with either meaning or values.

On the other hand, the biological spheres, including human society, cannot be detached from real life, and all environmental factors are related to human meaning and values. For this reason, we should not leave environmental problems to world of science. Instead, all stakeholders ought to join in the discussion of "environmental issues" from their own perspectives.

Though the word "issue" is also translated as "mondai" in Japanese, the problems cited above should also be seen as "environmental issues." Gluschkovim showed theoretically, at a time when the Internet had not yet become popular, ways in which computers could be used to enable up to 200 million people to discuss environmental issues (Moev and Gluschkovim, 1976). He also described how the referendum could be systematized as a way to allow issues to form, rather than people simply voting "yes or no." Recently in Japan nuclear power plants and/or sites for the disposal of solid waste have developed as issues, but it may be a greater problem that election campaigns have lost sight of them.

Differences between Analysis and Planning in Environmental Education

If we can manage a perfect information system, from the viewpoint of quantity, quality, computational speed, etc., the problems of the mistakes which prevail in the fields of science would disappear. As is often shown in a new field of "complex" science, however, a perfect information system has not yet been developed to solve problems. For instance, even though a huge computer can make paired comparisons of the merits and demerits of 10 planning variables in only 1 millisecond, an increase in the number of variables to 50 or so means that it would take more than 35 years to solve the problem with the same computer (Shiozawa, 1995). We should, therefore, make a distinction between analysis and planning.

Let us take the case of river pollution as an example of environmental analysis. After the pollution load measurement and subsequent observations of certain quality indices are performed, the data are then entered into a mass balance equation to evaluate the dispersion coefficient which represents the longitudinal mixing of this quality of material. If the quality is not improved simply by a pollution load cut-off, some other countermeasures will be needed, such as the introduction of water as a dilutant, or changing the stream configuration so as to vitalize its biotic potential, which will result in a change in the dispersion coefficient. But as procedures for the validation of the coefficients involved remain at the level of theory and have not been worked out in practice, methods for regulating river pollution have not been generally agreed on.

Strictly speaking, the methods of observation are no longer as simple as described above, so I will attempt to classify the new techniques roughly as follows:
1) the development of more precise physical or chemical techniques of quality analysis;
2) better methods of flow visualization to understand the hydrodynamic flow behavior; and
3) investment of large sums of money to allow the adoption of large scale modeling procedures.

In any case, generally agreed methods of observation now include results obtained by these techniques.

I will describe an actual case of eutrophication control in Lake Biwa or the Seto Inland Sea which was a major social issue in Japan in the 1970s. At that time it was said that the reduction of the phosphate load was an effective way to deal with the problem, but the environmental administration board's statement came after scientists announced that only part of the eutrophication process was fully understood. In July, 1980, Shiga Prefecture enacted an anti-eutrophication ordinance, pushed by an upsurge in citizens' anti-detergent campaigns. But formally, the board simply based their decision on a very simple laboratory experiment which showed that the red tide plankton proliferation was inhibited in a phosphate poor environment at a time when the most advanced research had already discovered and described the life cycle of the plankton and its cysts.

In my opinion, planning means the creation of social systems based on a sensitivity analysis of the range of constraining conditions through a system of morphologically autonomous rationality, involving all stakeholders as shown in Figure 1.

But here, it will be much easier to refer again to Gluschkovim (Moev and Gluschkovim, 1974) to characterize the planner or decision-maker. Important decisions for the future are made necessarily under some conditions of uncertainty. At this time, the job of analysts is to prepare data on what is the current social trend and what will occur following the decision. However, the analysts (mainly scholars or researchers) reveal their tendency to make things obscure. Even though the process of observation is endless, it cannot

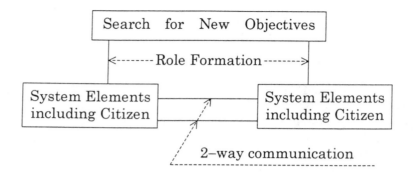

Figure 1. System of morphologically autonomous rationality

be allowed to prolong a decision endlessly. There will be an optimum moment for decision making. A leader who can choose the right time requires more than the ability to analyze, and should be trained accordingly. But the leaders of ordinary organizations spend all of their time concerned with small issues and cannot afford to think about future policy.

I would like to mention here one example of policy making in relation to risk management. When the Shiga Prefecture administration in 1983 proposed research on non-phosphorous LAS, which was not prohibited under the Lake Biwa Ordinance, the resident reporters at the Shiga Prefecture government whispered to each other that no one would be able to undertake such difficult research. They were surprised at the news that I had accepted the proposal. This is simple evidence that even journalists are living in a quasi-scientific world. My paper (Sueishi et al., 1988) seemed to be cited widely in developing countries where people suffer from both preindustrial and industrial chemical hazards.

Environmental Informatics and Language Skills

Attractive words like "environment," "international," "information," and "human" have been affixed to the names of many newly established universities and/or their faculties. If two of these are coupled, the name will be even more attractive, irrespective of the educational concept to which it refers. Although environmental informatics may be a typical case, this will involve quite important issues, because this field challenges the problem of literacy in environmental education. In this section, I give a brief critique of the common usage of this terminology.

Data on environmental quality, and especially on pollution, are the best known form of environmental information. The reason is that the administration is apt to conceal information either because it is unfavorable to the bureaucratic system or because the administration considers the citizens too sensitive. I have designated this bureaucratic control as the first stage of information dissemination (Sueishi, 1981).

The second stage of information dissemination is far more disorganized than the first because it involves researchers from disciplines with fixed boundaries which affect the way they define the environment.

Take the example of environmental economists. As a matter of course, economics deals with the economic environment and with perfect information systems which provide the ideal environment for *homo economics.* If it is assumed that economics is the study of human economic behavior driven by the imperfections of the information system, environmental informatics may be just the same as economics.

In this second stage, the ways in which economists can contribute to environmental information may include the conversion of a financial input-output analysis into an input-output analysis of chemical materials.

The reason is very clear. If we consider human activity which puts less of a burden on the environment, there may be important pollution emission points other than those involving air, water, and land. This idea is partly applied these days to the movement for "environmental housekeeping," but I propose that what we need is an environmental "fugacity" or dispersal model, based on economic input-output analysis, which takes into account the speed and degree of diffusion of materials into the environment, and their ability to affect it.

The third and final stage of information dissemination is really constituted by the informatic environment and it will definitely be concerned with issues of privacy, democracy, decision-making, etc. A more detailed discussion will be given in the next section, but I would like to mention here as an extension of the first and second stages the question of "language skills." I mean by these not simply rhetorical ability, but the ability to create new words to represent and communicate the environmental predicament.

When we talk about environmental affairs, we often encounter the meaningless usage of words such as "amenity," "comfort," "greenery," "quietness," or moralistic uses of words such as "waste." In the first stage of information dissemination, scientific jargon should not be used, and specialists must train their own language skills so that they can be understood outside their discipline in the second stage. When I proposed the new concept of "environmental capacity" (Sueishi, 1975) many researchers regarded this word as referring to the natural assimilation capacity. In the same way, Odum's term, "carrying capacity," was misunderstood by a famous professor of sanitary engineering in the USA to refer to pollution load transportation capacity in rivers. A lack of language skills seems not to be limited to the Japanese.

Paul Portney, president of RFF (Resources for the Future Foundation) gave a lecture titled "Risk, Public Policy, and SRA" at the SRA (Society for Risk Analysis) 1996 annual meeting (SRA, 1997) and put stress upon "language skills" related to risk communication, which include the ability to act as an advocate in such a way as to gain both credibility and people's votes. I will summarize these issues in the last section, in the discussion of the "new literacy" for environmental communication rather than environmental education.

A New Concept of Literacy — Environmental Communication

Even though a teacher insists that knowledge instruction is a necessary procedure to bring out students' abilities, current education cannot yet get away from "allonomy" or dependent thinking in which traditional social objectives are explicitly or implicitly presented as objective rationality in a top-down fashion. On the other hand, Figure 1 above shows a form of autonomous thinking based on the mutual communication between participants.

What, then, is communication? It will be easy to answer this question with words like "medium," "code," "addresser," and "addressee," but this answer does not signify the content which should be communicated. In a similar way, literacy has been defined as the ability to read and write, but it is not yet concerned with the content of what should be read or written. Speaking of the Internet, we should be aware of a new tendency for a "pipeline" industry, the information "superhighway," to be converted into a "contents" industry while the "citizen" is being replaced by the "netizen" (Kumon and Nishi, 1995).

Related to the discussion of content, Yamazaki(1997) heaves a sigh that the essential evils of Japanese journalism are attributed to poor language education in which too much importance has been placed on emotional sentences. He criticizes radically the Japanese people's lack of scientific literacy: in other words, "science in Japan is drowned in the sea of indifference of the general public." Noda (1996), who established the Association of Literature and Environment in 1994, describes Japanese literature as being mainly concerned with emotion and feeling, and suggests that research literature on environmental problems should be distinguished from other kinds of environmental literature which is merely emotional and sentimental.

It is said that a telephone cable is capable of 0.06 million bits of information a second compared with 1200 million bits per second using high vision television technology. Qualitatively speaking, the content of the latter at present consists mainly of the masses chatting, and not communication about the environment.

Concluding Remarks

Readers of this paper may possibly feel that I am a heretic. However, Tsuyoshi Sasada, an authority on computerized environmental literacy, has reported (1996) that the intellectual gap between professors and students is about to be disappear in Osaka University. I began research in 1977 at Osaka University to create intermediaries between the university and the region, an interim result of which was the setting up of a new system of citizen researchers at Suita City in 1992. Figure 2 which is reproduced from the USEPA manual for risk communication is in line with my scheme. Yes, citizens' investigations are the best way for the public to acquire environmental literacy and mediating power.

In 1983, I joined, as a Japanese delegate, the ACHEE (Asian Committee of Higher Environmental Education) for Planners and Decision-Makers in South-East Asia. It was proposed by Leandro A. Viloria (Institute of Environmental Planning, University of the Philippines) when the global environmental problem was not yet so hot. It is really regrettable that the Committee is now in suspension due to the lack of funds from UNESCO.

Totten (1997a) declared that Japan is a colony of the USA and he made the profound point that the reason is the absence of a true university in Japan: the Japanese imitation of a university should really be called a "professional school" (Totten 1997b).

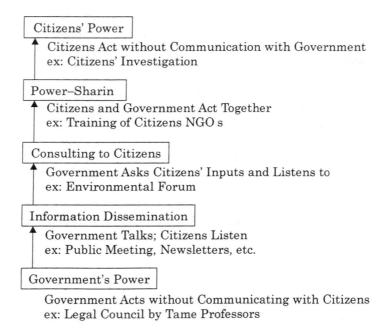

Figure 2. Simplified sketch of the ladder for citizens. Participation to environmental administration. (Source:US Environmental Protection Agency (USEPA),1987)

References

Kumon, S. and K. Nishi. 1995. Ima nani ga mondai nanoka: joho tsushin koshinkoku kara no dasshutsu sengen [What is the current problem?—Declaration of escape from an information-communicatively underdeveloped country], *Chuo Koron*, Dec., 1995: 108-114, in Japanese.

Moev, V.A. and V.M. Gluschkovim, 1976. *Konpuyta to shakaishugi* [The computer and socialism]. Translated from the Russian by Y.Tanaka. Iwanami Shinsho, Tokyo, Japan.

Noda, K. 1996. "Kankyo bungaku" nichibei shimpo no igi [Significance of Japan-US Symposium on Environmental Literature]. *Asahi Shimbun* (evening ed.), Oct. 1, 1996, in Japanese.

Sasada, T. 1996. Kankyo sekkei ni okeru virtual reality to shimin sanka [Virtual reality and citizen participation in environmental planning]. Seminar of IDEC, Institute of Democratic Education), Kinki Branch, Dec. 7, 1996, in Japanese.

Shimizu, I. 1972. *Rinrigaku noto* [Notes on ethics]. Iwanami, Tokyo, Japan, in Japanese.

Shiozawa, Y. 1995. Shakai system kenkyu no ichi shiten [A viewpoint on social systems research]. Open Lecture at Shiga University, October. 20, 1995, Otsu, Japan, in Japanese.

SRA. 1997. *RISK Newsletter.* 17(2): 1-4., Society for Risk Analysis, McLean, Virginia, USA.

Sueishi, T. 1975. Toshi kankyo no sosei: hakyoku kara no aojashin [The revival of the urban environment: blueprint from catastrophe]. Chuko Shinsho, Tokyo, Japan, in Japanese.

Sueishi, T. 1981. Toshi no joho kokai: shimin kara no atarashii shiten [Information dissemination in municipalities —A new citizens' viewpoint]. *Shisei Kenkyu* [Municipal research], 52: 42-53, in Japanese.

Sueishi, T., T. Morioka, H. Kaneko, M. Kusaka, S. Yagi and S. Chikami. 1988. Environmental Risk Assessment of Surfactants: Fate and Environmental Effects in Lake Biwa Basin. *Regulatory Toxicology and Pharmacology,* 8: 4-21.

Sueishi, T. 1990. Toshi mondai no doko to sono kankyoronteki taio [Trends in urban issues and corresponding environmental theory], *Kankyo Joho Kagaku* [Environmental Information Science] 11(1): 14-20, in Japanese.

Sueishi, T. 1997. Keikaku gyosei ni yoru risk management no kanosei to genkai [Possibilities and limits of risk management in planning administration]. *Keikaku Gyosei* [Planning administration]. 20(3): 15-20, in Japanese.

Totten, B. 1997a. Nihon wa Amerika no shokuminchi ka?! [Is Japan a US Colony?!], *Bungei Shunjyu,* May, 1997: 94-102, in Japanese

Totten, B. 1997b. Me wo samase, ohitoyosi no Nihonjin [Wake up, credulous Japanese], Lecture, Japan Techno-Economics Society, Osaka Branch, July 4, 1997, in Japanese.

Yamazaki, M. 1997. Kagaku to jihodo [Science and Journalism], *Asahi Shimbun* (evening ed.), July 31, 1997, in Japanese.

Chapter 15

Environmental Contamination and the Information Highway

Frank M. D'Itri, Institute of Water Research and Department of Fisheries and Wildlife, Michigan State University, East Lansing, Michigan 48823-5243, USA.

Introduction

Computer modeling projects represent some earlier technological adaptations to describe the environment. Modeling has achieved sophisticated levels and continues to improve with better software and new virtual reality programs. In the Institute for Water Research and Center for Remote Sensing at Michigan State University (MSU), long distance learning also began to be implemented at an early stage with new techniques to map the ground contours, soil qualities, and weather conditions. These had immediate application for farmers, and the MSU Extension Service was already in place to disseminate the information. The United States government sponsored extension service program had been charged with providing farmers with cutting edge technology for the last hundred years. However, the changes over the last two decades have been more radical than at any time previously.

In addition to sophisticated computer modeling and other display techniques, new communications hookups have made the reception of data possible virtually as soon as it was compiled. An additional impetus to adapt and communicate new learning technology was, of course, the research on pollution, first on industrial and municipal discharges and then on non-point source problems. Now that attention is being focused on contamination of ground water from agricultural sources, this issue more directly affects individual households. As more of them are equipped with computers, the opportunity is available for massive education programs of a scale and type not dreamed of even a few years ago.

While workshops, seminars, and educational programs such as the annual Agricultural Exposition at MSU have always attracted a substantial audience, the new computer communication systems will extend participation further. The Internet system was developed and supported by the US government initially to allow the military and civilian scientists to communicate with one another. Additional civilian applications have developed rapidly since the advent of the personal computer. Now local telephone hookups permit access to other computers on the Internet any place in the world.

The term, "information age," does not do justice to the possibilities of this emerging era in which anyone, located anywhere, at any time can be empowered as control of information moves from centralized systems to individuals. The immense potential is clear. The new learning systems are not just about hardware or software but about the wherewithal to change for the better the way we learn, communicate, and react (Zare, 1997). At the same time, those still struggling to overcome the fundamental barriers to knowledge: adequate nutrition, literacy, and a living wage are going to be even further isolated.

The Information Super Highway: How It Works

Computers are becoming more powerful and more portable, allowing access to more information at ever greater speeds and with more facility then ever before. The Internet can disseminate information throughout the world and reach audiences that have a variety of interests and agendas. In the 1990s improvements in the hardware capabilities and the ease of software use portended enormous growth in computers' role in science and society. Virtually every aspect of scientific research is impacted by computers, from e-mail and World Wide Web (WWW) based distribution of information to data collection, analysis and sharing, graphical display and simulation of data, reference searching and management, grant accounting, and maintenance of supplies and equipment (Bloom, 1996)

The Web can be used for educational purposes by: 1) accessing the information with a web directory or search engine; 2) as an integrated interface to distance learning; and 3) to supplement the traditional classroom rather than replace it. By integrating the vast quantities and types of information available on the Internet, the interactive capabilities of a web user and the multimodal nature of the web pages, the web can be a major instructional aid for conventional classes or for distance learning.

The World Wide Web, a client server on the Internet, is well suited for the transfer of information. It is comprised of a multitude of individual server computers that forward documents to Internet clients who access them with specialized browser software. Many web sites have search engines that scan through the site and/or other data bases for key words. Alta Vista, for example, is a single large source accessed by key words. Yahoo, on the other hand, is subdivided into areas such as: arts, business, education, government, health, science etc. Users select what is displayed, a kind of review process. Also, sending graphs or equations in an e-mail attachment should not be a problem in the near future as a method to "drag and drop" a graphic into the text is currently being developed.

As the World Wide Web has grown exponentially, maneuvering around it can be facilitated with a browser such as Netscape Navigator or Internet Explorer. They can access graphics, video, and audio files written in hypertext markup language (HTML) and display them on the computer screen. HTML is a computer-platform-independent method of encoding documents for presentation on the World Wide Web. This system applies tags to plain text to specify such parameters as structure, type sizes, and hyperlink addresses. Software provides the interface, called a browser, which enables users to click their way around the web and retrieve information from or interact with servers that hold HTML encoded documents. Such browsers have been the primary agents to drive the explosive growth of the web for delivery of environmental information. Underlying web technology, hyperlinks allow readers to move from one document or site to another with a mouse-click on a link embedded in the document, generally a word, phrase, or graphic. Documents stored in several computers make the information accessible to desktop and other client computers.

Browsers can also interpret graphic files encoded with GIF (graphics interchange format), and JPEG (joint photographic experts group) and incorporate them into the visual display on the screen. For other formats, the browser incorporates "helper applications" or independent programs launched in separate windows to open files. Recently, many programs have been developed as plug-ins that are independent but operate within the browser environment to provide integrated content.

Just when HTML is becoming familiar, VRML (virtual reality modeling language) is being introduced to allow the viewer to look at three-dimensional objects on the WWW. For many applications, HTML is a 2-D straitjacket. VRML, the three dimensional equivalent, was designed to be a universal description language that can be read by browser software on any platform. Virtual reality is the next step in the evolution of the computer from a series of vacuum tubes that represented binary states to graphics generators that create photo realistic images. With virtual reality the user is placed inside the generated image as it is assigned properties which make it seem real.

In other words, the user becomes a participant within the computational space (Bricken, 1990; Kreiger, 1996). Both HTML and VRML can be encoded with plain text that makes them independent of the computer program and enables locations or objects to be hyperlinked.

The sheer quantity of material can be overwhelming. One librarian at Michigan State University described the Internet as a "jungle" (Huey, 1996), and even a cursory effort to "surf the net" quickly demonstrates why. For example, if a researcher were interested in finding data related to the poisoning of human beings by organic mercury compounds during the 1950s and 1960s in Japan, an unlimited search initiated for <Minamata and Disease> would produce more than 241,000 hits on Alta Vista and 1,060,000 hits on Infoseek. When, however, <mercury poisoning> is entered as a single side by side key word entity, Alta Vista retrieved about 13,400 hits and Infoseek more than 160,000. From these results, two observations become apparent. First, the number of documents retrieved varies greatly from search engine to search engine; and, second, it is not practical to scan all of these citations, so even the preliminary search must be narrowed. This is most easily accomplished by including the appropriate limiting characteristics such as the Boolean and proximity operators as well as phrase and case sensitive requirements for each search engine. For example, searches on Alta Vista and Infoseek for <"Minamata Disease"> reduces the citations to 289 and 77 respectively. These numbers can be more readily screened and the more promising ones investigated further via the hyperlink option. Of course all numbers change as materials are added.

The advantages of these types of electronic library resources lie, first of all, in the fast searches of electronic databases as a function of subject headings, title, and author. Abstracts and some full text articles can be downloaded to desktop computers. For broader topics, however, as the examples indicate, unless the search is closely limited with appropriate operators and highly selective key words, more citations may be identified than can reasonably be evaluated. Another major disadvantage is that most of the pre-1970 papers, reports, and abstracts are not in the electronic databases. In addition, many of the earlier state and federal databases such as STORET environmental data systems are random data sets that have not been evaluated with respect to validity or statistical confidence limits.

Among the important considerations that may be blurred on the internet is the distinction between peer reviewed and/non-peer reviewed literature. They were usually separate and distinct in paper publications. Articles were carefully reviewed by experts in each field to insure that the information was accurate and complete before publication. However, the process was slow. In the United States often more than six months passed between submitting a manuscript to a journal and its publication. However, the reader could be reassured that the methods, data, and conclusions were scientifically responsible and acceptable. In contrast, with an information search on WWW the distinction between peer reviewed and non-peer reviewed literature can be blurred depending on the source and method of access.

Being familiar with the individual researcher's standards offers some degree of quality control, but the usefulness of the information still has to be evaluated. How can the merit, accuracy, or validity of individual citations be evaluated unless the author or the citation appears in a peer reviewed journal? Quite often the web searcher must make an initial value judgment on a citation's usefulness based on its title and whether it would require excessive time to either obtain the journal article or additional information from other sources to evaluate it.

Electronic Database Collections/Resources

Disciplined and systematic process research entails the collection of information to draw a conclusion that adds to knowledge. The initial step in any research project is to search the literature for prior knowledge and to continue the search to add new knowledge. With the explosion of scientific and social research and new technology since the mid-Twentieth Century, the ever greater volumes of data generated and published have made it very difficult for researchers to remain current in their fields. They used to rely on traditional library accessing processes such as review articles, abstracting services, and professional journals. With the advent of the Internet, a source can be sought on a desktop PC much more readily, rapidly, and extensively; but then they have to be evaluated.

Electronic reference computers that access on-line information databases are an integral part of modern libraries. The information resources available through on-line systems are very extensive and provide a method to obtain specific, subject related, sometimes peer reviewed information. The scope and depth of these resources provide rapid access to an extensive electronic collection of bibliographic, abstract, and full text information, much of it from books, magazines, and newspapers.

The World Wide Web hypertext-based technology (HTTP) accesses a wealth of information, much of it housed at university and college libraries worldwide. At Michigan State University subject guides to internet-accessible information indexes and abstracts contain approximately 1550 scholarly and general interest titles, the most current 6 months of the New York Times, as well as full text or full image coverage of over 500 titles. In addition, the Michigan State University Library <http:// www.lib.msu.edu> accesses "First Search," an online reference service with 60 reference databases, including the full text of a million newspaper and magazine articles, citations from 30 million books, and abstracts of articles from more than 15,000 journals and newspapers. The "ISI Citation Databases" for science, social science, and arts and humanities are available from 1987 to the present.

The patron can not only conduct topic and author searches but also find articles which cite these works or related references. Academic journals are gradually being converted from print to the electronic format. As of the end of 1996 almost 1000 journals were available on the Internet. A number of important environmental journals are available on a commercial basis. The advantage of the electronic journal lies in its greater searchability and accessibility. For example, ChemCenter <http://www.chemcenter.org> brings together at one web site the extensive on-line resources of the American Chemical Society. Some problems that slow the implementation of electronic journals include: the peer review process, copyright, database sharing, financial support of electronic media, and the unwillingness of some authors to publish electronically.

In addition, a wide assortment of unorganized, non-peer reviewed data, opinions, and anecdotal information, not to mention out and out fantasy are available on the web. While publication is unrestricted and can be anonymous, on-line it is also instantly available to a very large audience. This disadvantage is that systematic professional evaluation through peer review is short circuited. How serious this tradeoff is remains to be seen. Meanwhile, the tradeoff in defined quality is a substantial increase in the quantity of materials available as we ride the information superhighway.

Desktop Video Conferencing

In the past people had to interact in person with their cohorts to communicate visually at an appropriate professional level. Desktop video teleconferencing can offer a real time "face-to-face" meeting among participants in locations as close as within the same university or company or as far away as halfway around the world. This can be very economical because video conferencing integrates the power and effectiveness of face-to-face communication without the time and expense of traveling to bring presenters and audiences together.

Video conferencing can leverage an organization's assets of people and information in new and innovative ways. Instead of spending time and resources traveling to and from a meeting, not to mention suffering travel fatigue, video conferencing allows the most knowledgeable experts to present their information around the world without leaving their office.

The viewers can switch among different presentation media, such as video tape, a computer based screen, a three dimensional item on a document camera, a shared white board application, and live shots with audio feed of the speaker and other participants. Such multiple options should enhance participation although some confusion and inattention may result as well.

Although desktop video conferencing is in its infancy, the World Wide Web is likely to play a major role in advancing this technology over the next 4 to 5 years. A major disadvantage is bandwidth limitations. This is likely to change as cable and direct broadcast satellite companies offer high bandwidth Internet access in the next few years. An integrated multimedia communication network will be able to deliver voice, data, and video at full speed in real time. Currently, the desktop video conferencing system is composed of a PC, hardware and software, a small camera, and either a digital or low speed analog phone connection. A small video conferencing window is displayed on each participant's computer screen. An advantage of this system is the ease with which spreadsheets, word processing documents, and/or web pages can be shared with other video conferencing partners.

Environmentalists use the web as a forum to express opinions and participate in discussions as well as to develop collaborative research proposals. Examples of such forums are the Environmental News Network (ENN) accessed at <http://www.enn.com/>; the Centre for the Study of Environmental Change <http://www.susx.ac.uk/Units/gec/gewfront.htm>, and the Environmental Organization Web Directory, <http:// webdirectory.com>. These web sites offer environmental scientists and concerned citizens throughout the world a format in which to conduct a two-way dialogue not only to obtain information but also to alert readers and inform distant environmental action groups about emerging problems. The advantage and disadvantage of the chat rooms, bulletin boards, and/or academic forums is that they can be anonymous. Therefore, while information can be exchanged rapidly to help solve common problems, the participants may elect to remain anonymous.

Despite some murky aspects, the Internet can be a very effective method to obtain information in well defined and emerging fields. For example, Lawrence Doe (1997) read in Edward Tenner's 1996 book, *Why Things Bite Back: Technology and the Revenge of Unintended Consequences*, that during the 1980s many IBM computer tape drives kept inexplicably crashing all across the United States and no one could figure out why. Tenner reported that a team of chemists at the IBM Almaden Research Center finally discovered that these computers were suffering indirectly from Legionnaire's disease. After the Legionnaire's outbreak in Philadelphia, PA, many public buildings added bactericides to their air-conditioning systems. These compounds contained tiny traces of tin. The bactericide containing solution vaporized and was carried through building vents so the tin was deposited on the heads of IBM tape drives, causing them to crash. Doe wanted to verify the story but found no mention of it in the printed literature, so he went on-line to ask if anyone knew the details. The very next day he had a message from the scientist who had found that link.

As this example demonstrates, the World Wide Web facilitates communication and reduces isolation. For many environmentally related self-help topics, one place to start is the Training and Development Resource Center <http:// www.tcm.com/trdev/>. It refers to itself as a "virtual gold mine of resources for the training and distance education on line community." Some of the educational items on this site include:

1) a training and development business section with hyperlinks to nearly 50 suppliers of training products and services;
2) a bookstore with titles that cover such subjects as career development, learning organizations, and mentoring;
3) a job bank to post resumes and search for available positions;
4) a professional's toolbox that identifies handy links to sites for news, internet information, search tools, travel aids and information, personal development, and reference guides;
5) a list of internet browser-based news groups that discuss training issues; and an extensive list of nearly 30 email based discussion list servers for adult education, distance learning, instructional technology, and much more. (Knotts,1997)

Other web sites provide a wide assortment of information on environmentally friendly technologies, data, legislation, training manuals, and online training. Among these are:
1) The UNEP International Environmental Technology Center <http:// www.unep. or.jp/>;
2) The United States Environmental Protection Agency <http://www.epa.gov>;
3) The United States Department of Commerce National Oceanic and Atmospheric Administration <http://www.noaa.gov/>;
4) The Consortium for Earth Science Information Network <http://www.ciesin.org/>;
5) The Great Lakes Information Network <http://www.greatlakes.net/>; and
6) The United States Department of Energy <http://www.doe.gov/>.

The World Wide Web is also useful to access government regulations and track pending legislation through the system. Industry organizations such as the American Automobile Manufactures Association, <http://www.aama.com/>, Chemical Manufacturers Association, <http://www.cmahq.com/>, and the Synthetic Organic Chemicals Manufacturing Association, <http://www.socma.com/ services.html>, can also be accessed for their position papers. Waste water topics include: the operation and maintenance of treatment plants, industrial and pretreatment courses, advanced waste treatment, and treatment of metal waste streams. Drinking water topics include: water treatment plant operation, small water system operation and maintenance, and water distribution system operation and maintenance. (Knotts,1977)

Another source, Cambridgesoft, is a free chemical database searchable over the World Wide Web, <http://chemfinder.camsoft.com/>. Chemical information has been collected from all over the world, and links are provided for what it terms "Yahoo of Chemistry." More than 3500 compounds are searchable by name, by chemical formula, and by substructure. The web site includes links to more than 30 sites with relevant information, such as EPA Toxics, Material Safety, Data Sheets archive at the University of Utah and the Texas Clean Air Act.

University web sites also offer a wide variety of information. For example, the web can access information resources such as the Michigan State University Extension bulletins at <http://

www.msue.msu.edu/>. For distance learning courses on drinking water and waste water treatment operation and maintenance, one helpful site is the Office of Water Programs at California State University, Sacramento (CSUS) at <http://www.owp.csus.edu/>. This site focuses on correspondence training materials developed for the US. Environmental Protection Agency (USEPA) and administered and monitored by CSUS. Students study the materials at their own pace and return their work to be graded. Courses can usually be completed within 50 to 150 hours of study. Continuing education units and certificates of completion are awarded when the course work is completed, and university credit can also be obtained.

Another useful distance learning opportunity is the National Environmental Training Center for Small Communities (NETCSC) at West Virginia University, Morgantown, WV <http://www.estd.wvu.edu/netc/>. Also funded by USEPA, NETCSC provides training to help environmental professionals develop skills and knowledge so they can better plan, manage, and operate drinking water, waste water, and solid waste systems in small communities. A special feature of this site is their training discussion forum. It lets the user network with other trainers and technical assistance providers.

Pollution Prevention, an Industrial Application

Aggressive, shortsighted development practices are endangering the world future generations will inherit. Instead of sustainability, deterioration of the environment more often is exhibited. The strength of the web is that environmental issues can be brought to the attention of a growing number of people. They can be empowered with information and problem solving methods to remedy their current problems and plan to avoid future environmental disasters.

As with applications for citizens, industrial and business manufacturing applications are innumerable, and much information can be obtained on the internet. One source is the National Pollution Prevention Center web site at <http://www.snre.umich.edu/nppc>. It included hyperlinks to other related World Wide Web sites. The web facilitates the exchange of information related to recyclable commodities, chemicals, and byproducts as well as used and surplus items or materials. The sites are numerous and variable, ranging from free access public sites dedicated to information on recycling such as the Global Recycling Network, <http://grm.com/grm/home.htm>, and Recycler's World http://recycle.net. More specific organizations such as ChemConnect, <http://www.chemconnect.com>, ChemicalWeb, <http://www.chemiweb.com>, ChemNet,

<http:// www.chemnet.com>, ChemTrade, <http://chemtrade.com>, First Chemical Market, <http:// www.2wchemnet.com>, and WWW Chemicals <http://www.chem.com>, offer a comprehensive directory of "wanted" and "for sale" listings for chemicals, chemical equipment, and laboratory instruments.

Information on recycling in the automotive industry is available from the American Automobile Manufacturers Association at <http://aama.com/environment/overview.html> and through its hyperlink connections to Ford, Chrysler, and General Motors web sites. Recycling scrapped vehicles reduces the solid waste stream. Presently, in the United States about 10 million vehicles, 94% of those scrapped, are collected for reuse and recycling. At least 75% of the steel, aluminum, copper, and plastic is recycled, some of it in the manufacture of parts for new automobiles. While companies have been hesitant to exchange trade secrets on production, they are eager to learn from each other about methods to reduce waste as well as the cost savings.

Environmental Leadership Training

The Leadership for Environment and Development (LEAD) program sponsored by the Rockefeller Foundation (1996), maintains an ongoing Internet dialogue through LEAD International Inc. accessed at <http://www.lead.org/>. This global electronic computer communications and information system enables the LEAD associates, fellows, and faculty to exchange data. This promotes multi-cultural and interdisciplinary collaborations that will contribute to the growth and support of sustainable development options for the future.

Conclusion

Now it is difficult to believe personal computers were a new toy less than fifteen years ago. We have to assume that the uses found for them so far are only the beginning. Clearly, the collection and dissemination of information via the Internet will increase greatly in the future. Already, we are compelled to rethink how we approach information gathering and dissemination, the impact on environmental issues and, most of all, new configurations that will affect how future generations are educated and motivated to demand and implement changes to achieve a healthy, sustainable environment.

References

Bloom, F.E. 1996. The road to staff competition. *Science* 273: 559.

Bricken, M. 1990. *Learning in Virtual Reality. Human Interface Technology Laboratory*, Technical Publication No. H1TL-M-90-5, University of Washington, Seattle, WA 98195, USA.

Doe, L. 1997. Internet comes through. *Chemical and Engineering News*. July 21 (87):48.

Huey, T. W. 1997. Personal Communication, Michigan State University Library, East Lansing, Michigan, USA.

Knotts, J. 1997. Internet training sources: let your fingers do the Surfing. *E-Train* 6(4): 18.

Krieger, J.H. 1996. Doing chemistry in a virtual world. *Chemical and Engineering News*, December 9, 1996 pp 35-41.

Rockefeller Foundation. 1996. *Global Environment. The Rockefeller Annual Report*, New York, NY, USA, pp 47-49.

Tenner, E. 1996. *Why Things Bite Back: Technology and the Revenge of Unintended Consequences*, Vantage Books, New York, NY, USA.

Zare, R.N. 1997. Knowledge and distributed intelligence. *Science* 275:1047.

Chapter 16

Networks, Technology, Distance Learning and Environmental Education

Ed Brumby, Learning Resources Centre, Deakin University, 221 Burwood Highway, Burwood Victoria 3125, Australia.

Introduction

Our dealings with, management of, and teaching and learning about our natural environment are marked by a number of paradoxes. If, as the Kenyan proverb suggests, we do not own the land but simply hold it in trust for our children, why is it that a fundamental aspect of environmental education is to take note of and learn from the multitude of mistakes that we have made and continue to make?

Is it simply that we have failed to take that in-trust commitment seriously enough? Or is it because, in our pursuit of social, technological, and economic progress, we have become so far removed, physically and spiritually, from our natural environment that we have lost cognizance of that trust? And herein lies another paradox: the systems and technologies that have served to distance us so much from our natural environments may yet serve to provide renewed awareness and understanding of the preciousness of those environments and revive our commitment to their maintenance and sustenance — for the sake of our children.

Networks(1): People and Institutions

In a general sense, networks provide the means to exchange information, ideas, knowledge, and learning. They represent, in a human sense, people communicating, cooperating, collaborating, and learning from each other to achieve common goals.

Networks of people form the basis of local communities, neighborhoods, villages, and towns. As local and community institutions, schools and universities belong also to national, international, and global networks and communities of learning. Governments and their various arms and institutions form their own networks but are also a fundamental component of local, prefectural, regional (in-country), and national networks, and are members of other regional and global government-to-government networks.

Business and industry operate in similar fashion. All of these: local communities and neighborhoods, schools and universities, government, business, and industry have their own interconnected roles to play in caring for, managing, and teaching and learning about the natural environment — supported by, and in concert with those non-government organizations, voluntary groups, and other communities who have a special interest in environmental protection, management, and education.

Sudara (see Chapter 12) provides several illustrations of the efficacy of using local community and volunteer networks not just to manage and care for the environment, but to play an activist role in protecting it from misguided government and private enterprise development projects and, in so doing, influence contemporary and future environmental management policy and practice. He cites the work and influence of Buddhist monks in Thailand (as pervasive and influential a network as can be found anywhere) in protecting individual significant trees and how this has been extended by community volunteer groups in various parts of Thailand to afford protection to whole forests. He also describes the fruitful connections that have been made with, and the networks established between individual communities and government organizations and authorities through cooperative research and information-gathering and dissemination activities associated with the protection and maintenance of coastal environments.

Toda and Matsui (Chapter 19) also describe how individuals and households can work with government agencies to collect and analyze environmental data — in this case, meteorological data in the Biwako area of Japan. Of significance in this report is that computer and communications technologies are being used to provide: 1) greater speed in data collection and transmission (versus the slower, more traditional approaches of written, mailed reports still used in many parts of the world), 2) greater power to provide different kinds of data in different media, and 3) greater incentives for individuals to become involved in environmental matters and, in turn, to stimulate greater interest within the community generally about environmental issues.

Le Grew (Chapter 18) reminds us that environmental problems are not constrained by geopolitical boundaries and that there are opportunities, if not imperatives, for the network of nation-states of the Asian-Pacific region to work collaboratively and to take a distinctively regional approach in attending to the many and common environmental problems that beset so many countries in the region, and to environmental management, education, and research in general. He proposes the establishment of an "Enterprise Web" (an extended network, if you like) involving the cooperation of governments and their agencies, business and industry, educational institutions, and local communities as one means of disseminating and implementing this approach.

Networks(2): Technologies

If networks in general provide the means for the exchange of information, ideas, knowledge, and learning, the new technologies provide the means to do it in a faster, more powerful fashion. Computer technologies enable the storage, analysis, manipulation, and presentation, in many different ways and using different media, of vast amounts of data. Connection to the Internet provides almost instant access to this data from anywhere in the world, regardless of time or place. Other recent developments in telecommunications mean that access to telephones for the transmission and receipt of information is no longer a significant difficulty, even in the most remote parts of the world. Videoconferencing with terrestrial or satellite networks has become part and parcel of the everyday operations of government, business and industry, and educators on every continent. The opportunities afforded by these technologies to those who have any interest in the natural environment are limited only by the individual and collective imagination.

The use of technology in the collection and analysis of atmospheric data is referred to below (see Toda and Matsui, Chapter 19). Sakamoto (Chapter 17) describes how Japan, like so many countries in the Asian-Pacific region, has embarked on a program to ensure that all schools are provided with resources and facilities to enable them to access the Internet. He also outlines the innovative use of satellite delivered videoconferencing to link universities throughout Japan and, through other telecommunications networks, to their sister institutions throughout the world.

Le Grew (Chapter 18) proposes the utilization of the Internet and associated technologies to establish an Environmental Education Network (EENET) — an international network of university-based web sites providing access to electronic warehouses of environmental information, communications (email, computer-conferencing etc.), and other electronic services and facilities for educators, researchers, and anyone else who is interested in environmental matters — as the first step in establishing the Enterprise Web referred to previously.

Networks(3): Education

If, as the Kenyan proverb cited earlier suggests, we do not own the land but simply hold it in trust for our children, we must exercise and act upon two important responsibilities: to care for and nurture the environment, and to teach our children to do likewise. A third major responsibility is implicit in the first and has already been alluded to: to continue to learn as much as we can about our natural environment and to share our growing knowledge and understanding so that we can all learn from each other.

The human and institutional networks already in place and described earlier provide the means and wherewithal to facilitate and foster this learning. The challenge to teach our children about, and to foster greater understanding of our environment is being taken up, as described elsewhere in this volume, by governments and their agencies, schools, and community and voluntary groups everywhere. Sakamoto (see Chapter 17) describes briefly some of the environmental enterprises and activities being undertaken in Japanese schools and goes on to explain how access to the new communications technologies will enhance these. He also emphasizes the need to not only provide access to such technologies, but to provide training in their use and effective application.

In a similar vein, Hirai (Chapter 20) describes how teachers and students at Shiga University are using flexible/distance learning approaches as an innovative means of enabling students to use and apply computer and communications technologies. Le Grew's EENET proposal is based on utilizing and exploiting the capacity and power of these information systems to provide access to, and to facilitate learning, research, and the sharing of knowledge in a cyber-environment where time and place no longer act as constraints in the search for understanding.

One institutional "network" that appears to be missing in all of the papers summarized here is the business-industry-private enterprise network. While there is evidence to suggest that many business enterprises are taking a more enlightened approach to dealing with and protecting the environment, there is considerable scope for them to play a greater role in supporting, facilitating, and financing the kinds of educational and other cooperative programs and projects described herein. It is simply not enough to rely on governments to respond to, or take the initiative entirely in fostering effective environmental education programs, locally, nationally, or internationally and to provide the resources and infrastructure to do so. Business and industry contribute, more than any other group or network, to many of the environmental problems that plague the Asian-Pacific region. They have a clear responsibility to play their part as members of the human network in ensuring that we leave to our children not just an environment that is healthy and flourishing, but the means to acquire greater understanding of how to ensure that they leave it in even better condition to their children.

Chapter 17

The Use of Advanced Information and Communication Technologies in Education

Takashi Sakamoto, National Institute of Multimedia Education, 2-12 Wakaba, Mihama-ku, Chiba 216-0014, Japan.

Reports by the Central Council for Education
"Zest for Living "(Ikiruchikara)

Information technology and environmental education are the most important problems in educational innovation in the run-up to the 21st Century. Since its introduction in a report by the 15th Central Council for Education on July 19, 1996, the term "zest for living" seems to have become widely used in the educational community.
"Zest for living" is believed to comprise the following two elements:
  1) the ability to identify problems for oneself, learn by oneself, think for oneself, make independent judgments, act for oneself, and be able to solve problems; and
  2) a rich sense of humanity capable of self-reliance, cooperation with others, compassion for others, emotional sensitivity, and the health and physical strength required to live a vigorous life.

In global terms, education in Japan over the past fifty years from the end of the Second World War can be considered quite successful. Nevertheless, the present state of education raises many problems. With the 21st Century approaching we are on the verge of developing a fully-fledged advanced information society. In turn, we will need to train people able to think independently, and able to develop society even further. The basic human ability to meet to these two requirements is thought to be embodied in what is referred to in educational circles as "zest for living." Imbued with this quality, the response of children to torments and bullying by their classmates will not be suicide but perseverance and survival. Positively, the term "zest for living" is more aptly described as "the strength to survive."

"Room to grow" (yutori) is believed to be a prerequisite for fostering this "zest for living." To this end, the Council report called for a reduction and review of the excessive time devoted to specific educational subjects during school so that schools can develop children's zest for living in collaboration with the members of their families and local communities who share responsibility for education. The report continued that schools should use the hours gained to provide a "period for integrated study," mainly to cover educational subjects related to the 21st Century.

The 16th Central Council for Education

The current 16th Central Council for Education presented a second report on June 26, 1997. The report has four proposals.

The first proposal is for continuous education from junior to senior high schools. In order to diversify secondary education and provide children with the "room to grow," the Council proposed the introduction into public schools on an optional basis of a continuous education system from junior to senior high school. Under this system a high school entrance examination would no longer be needed. If such a system were adopted, it would enable the adoption of a curriculum relevant to the local community, focusing on international issues, information technology, and the environment, as well as of practical skills and training, the transmission of traditional skills, and the teaching of slow learners.

The second proposal is a reform of high school and university entrance examinations. The Council recommended that the present-day knowledge-oriented examinations to test students' scholastic ability in selecting entrants should be reformed through the adoption of a wider variety of examination procedures and the use of multi-dimensional measures to evaluate zest for living. The basic academic ability acquired at high school can be assessed by the examination conducted by the National Center for University Entrance Examinations. Under this type of system, students would be encouraged to take examinations on as many subjects as possible, while individual universities can use interviews, essays, and tests to measure specific skills.

The third proposal is to allow students to skip a grade to enter university quickly in some circumstances. The proposed scheme would let students with exceptional talents in mathematics or physics enter university directly from the second grade of high school on the recommendation of their high schools.

The fourth proposal is for an education system that meets the needs of the aging society. Under this educational system, schools would teach students about the characteristics and problems of the elderly so that they can develop both respect and compassion towards their seniors, and a general motivation to serve others. Accordingly, students should be provided with the opportunity to learn from the experience of their elders, and the elderly should be provided with lifelong education and support.

Information and education

Most of the recommendations concerning science and technology education were included in the first report. First of all, the most important issues in the response of education to social change are related to information technology. The recommendations were specific and included the increase of information literacy for the advanced information and communication society; the promotion of education in information methodologies and systems throughout elementary, junior high, and senior high schools; the linkage of all schools to the Internet; and the development of "open schools" active in transmitting their own information.

Development of education in science and technology

With respect to science and technology the report addressed the need to emphasize children's ability to think independently and to look at things in a scientific way and to learn through experience. Also addressed is the importance of learning processes which enable children to feel motivated, ask questions, and make inferences, as well as the importance of letting them experience the joy of discovery and creation through repeated trial and error. Further, the number of mandatory subjects will be carefully reduced to allow more flexible use of time. Teaching environments are proposed which utilize, for example, team teaching, human resources in the local community, visits to science museums and businesses, and the provision of information on learning opportunities.

Environmental problems and education

In order to improve and strengthen environmental education, it is important to promote it from the point of view of "learning from the environment," "learning about the environment," and "learning for the environment." It is also important to promote environmental education in ways which are relevant to other academic subjects, ethics, and special activities, and which attach importance to active learning.

We must nurture concern for the environment and nature and cultivate the attitudes, temperament, and ability to act independently to preserve and improve them. It is desirable to enhance the teaching ability of teachers through the expansion of initial and in-service training, and to make use of nature conservationists and other members of society in training programs.

In order to improve environmental education through the provision of a variety of learning opportunities in the local community, it is important to expand opportunities for children to observe nature and to get to know about it in other ways, and to expand opportunities for various forms of active learning such as environmental study classes at national children's centers. It is also desirable to encourage universities and companies to help arrange opportunities to study environmental problems, to improve the provision of information on the various study opportunities available, and to encourage volunteer activities related to environmental problems from the standpoint of what each person can do in their own immediate surroundings.

Basics and Fundamentals of Scholastic Ability in the 21st Century

A system for the rigorous selection of specific items for education in information systems, science, and technology is under consideration at the Curriculum Council, an advisory body to the Japanese Ministry of Education, Science and Culture. Initially it is desirable that the teaching curriculum be restructured. Ideally people should not conclude that because reading, writing, and arithmetic are the traditional basics in education they are the only subjects that require emphasis. With the advent of the advanced information and communication society new forms of reading, writing, and mathematics are required. More specifically, the reading and writing of the 21st Century will be concerned with expression and communication, while mathematics will be concerned with logical thinking.

Utilization of the "Period for Integrated Study"

The "Period for Integrated Study," which will be made possible through the more precise selection of educational subjects, will be allocated to types of education relevant to the advancement of modern society. This education encompasses subjects such as information systems, international understanding, and environmental studies. One easy way to do this is to give information systems a central role in the teaching of international understanding and environmental issues. Children will become motivated to study international communications and the environment through: 1) an understanding of foreign languages and cultures aided by multimedia technology and information networks; and 2) an understanding of the environment through activities in which they can express themselves using multimedia technologies.

To implement this successfully, it is desirable to establish separate curricula for international understanding, informatics, and environmental studies. We need then to consider what items in those curriculum should be integrated and allocated to which grade according to the children's stages of development. To implement teaching and learning focused on education in information systems effectively, schools must take the following steps:
1) implementation of integrated studies programs;
2) tie-ups with local communities to makes schools more open;
3) utilization of external information with the help of multimedia technology and the Internet;
4) the hiring and use of computer programmers and special part-time instructors; and
5) appropriation of the necessary budget.

Also, with respect to educational administration, efforts need to be made in relation to:
1) multimedia technology;
2) the enhancement of teaching materials and experimental and training facilities and equipment;
3) the provision of databases of teaching materials, and other educational information;
4) improvement of the Internet; and
5) the training of teachers to teach the integrated studies programs.

Educational Policies Associated with Educational Reform

There have been many proposals for educational reform by organizations other than the Central Council for Education. These include:
1) January, 1995: Ministry of Education, "Promotion of educational policies adapted to the development of multimedia technology." (Panel summary);
2) February, 1995: Headquarters for the Promotion of the Advanced Information and Communication Society, headed by the Prime Minister, "A Basic Policy for the Promotion of the Advanced Information and Communication Society;"

3) August ,1995: Ministry of Education, "Guidelines on the Adaptation of Information Technology in Education, Science, Culture and Sports;"
4) November, 1995: "Science and Technology Basic Law;"
5) July, 1996: "Science and Technology Basic Plan;"
6) July, 1996: Ministry of Education "Task Force on Utilizing Multimedia in the Higher Education of the 21st Century;"
7) January 24, 1997, revised in August, 1997; Ministry of Education, "Educational Reform Program;" and
8) May 16, 1997: Cabinet decision on the "Action Plan for the Reform and Development of the Structure of the Economy."

All of these studies and plans discuss extensively the issues of the introduction of information systems into education and the promotion of science and technology education.

Utilization of the Internet and Satellite Communication

In May, 1997, the Cabinet adopted the following: "The improvement of the ability of every citizen of Japan to make use of information systems is an indispensable factor in maintaining and increasing Japan's economic vitality and international competitiveness. The provision of an information and communication network at the level of the schools and the active use of information and communication technologies enable children to learn beyond the physical constraints of classroom teaching, thereby fostering their creativity and their ability to think and express themselves, and to realize their tremendous potential."

Concerning elementary and secondary education, the same cabinet decision holds that:

1) educational centers, schools, and other institutions across the nation should be linked in a common network in the near future to provide schools with Internet access and other on-line resources;
2) instruction that makes effective use of computers and networks should be reinforced, through cooperation with local and overseas communities; and
3) hardware and software geared to support multimedia-aided creative activities should be upgraded or enhanced.

Through this decision, the introduction of the Internet into schools has now been adopted as a national policy. In preparation, the so-called "100-school Networking Project" and the "KONET Plan" have also been carried out. In June, 1994, the Ministry of International Trade and Industry announced its advanced information program, designed to compete with the U.S. National Information Infrastructure Initiative, together with plans to promote an information oriented society in cooperation with other Japanese ministries and agencies. One such plan was the 100-school Networking Project, and to promote the "advanced utilization of information through the networking of schools." Since 1995, 111 schools selected from among the 1,543 that applied to participate in the project have been working on novel uses of the Internet for the collection, dispatch, and exchange of information; for collaborative learning; and in the education of disabled children.

Under the KONET Plan established in November, 1996, a total of 1,014 schools were linked with the Internet to facilitate interaction and information collection via email, and support was given for the creation and dispatch of WWW home pages for a number of educational applications.

Primary school children cooperate with children in other countries such as America, Australia, and Canada in projects related to global warming, the measurement of acid rain, natural observation, the measurement of deforestation, and so on.

In fiscal 1997, the Ministry of Education's budget specified twenty-five regions nationwide in which practical research projects for the study of specific topics in subject learning, collaborative learning, and international communication could be launched, all of which make use of the Internet. Educational centers and other institutions will be widely linked with schools in these regions in an information and communication network geared to improving the learning environment and enabling schools to obtain educational information via the Internet. Activities in 15 regions started in 1997.

An attempt to undertake distance education by connecting schools in remote areas to the schools in the cities by fiber optics is being undertaken in seventeen places throughout the country. Since fiscal 1996, joint teaching via communications satellites has been started by teams linking the Aogashima Elementary School and the Joto Elementary School in Chuo Ward in Tokyo as well as the Kita-Daito Elementary School in Okinawa and the Maeda Elementary School in Urazoe City. In addition, seven teams made up of general high schools located in the mountains have paired up with specialized high schools in the cities. In fiscal 1997, classrooms in hospitals were due to be connected with their parent schools by way of optical communications.

For in-house training in enterprises the use of satellites is not rare. Recently, teacher training and social education through ISDN (Integrated Service of Digital Network) and communication satellites has also started.

In higher education, which previously had been lagging behind in these types of developments, the full-fledged use of a communications satellite was started under the Space Collaboration System Project, a new project designed primarily for national institutions of higher education. The system provides interactive communications through audiovisual images and allows easy transmission of instructional and other materials created on video, image-display equipment, TV cameras, computers, etc. This has helped enhance effective information interchange, thus bringing university education into the space age. By the end of this fiscal year, a total of 89 VSAT (Very Small Aperture Terminal) stations will be set up in over half of Japan's national universities (55 out of 98), together with eight technical colleges, and 10 research institutes.

Satellite-aided continuing education ("refresh kyoiku") is also in operation. Some of Japan's private universities provide higher education nationwide by using communications satellites. We may well say that the multimedia age has arrived at schools.

Research and Development of New Educational Subjects

In a measure geared towards curriculum development, the Liaison Committee on Science Education of the Science Council of Japan brought together academic societies representing mathematics, science, technology, information, and educational technology together with the members of the Fourth and Fifth Divisions of the Council. In June, 1996, they published their "Proposal for New Curriculum Development Oriented Towards the 21st Century." Going beyond the proposals of individual academic societies, this proposal is expected to become a benchmark for curriculum development in future. The Liaison Committee also compiled a report on "Teacher Education for the 21st Century." A subcommittee on new curriculum development studied the curricula for comprehensive science education and published a book entitled "Proposal for a Curriculum Suitable for Education in the 21st Century."

The project to research and develop high quality software for learning is also producing results. The project funds teachers from universities and elementary, junior high, and high schools to help them develop high-quality software for learning in collaboration with educational software developers. With ¥20 million ($0.2 million) granted per development unit under this project, the sums awarded so far have amounted to ¥200m in Fiscal 1994, ¥340m in 1995, and ¥400m in both 1996 and 1997, respectively. The software developed has been commercialized, and some programs have received the Nihon Software Grand Prize and the Award for Excellence.

Science and Technology Education at School

In parallel with these efforts to improve the educational environment, it is agreed by many that science teaching at elementary and junior high schools should not only focus on learning through experience, e.g., observations and experiments, but also that natural phenomena, nature-human relationships, nature-society relationships, and other topics should be taught through special activities, school events, and other subjects such as Japanese, social studies, drawing and manual arts, music, physical exercise, home economics, information, international relations, environment, security, and health. Observations, research, and experiments should be dealt with in the form of team teaching and as comprehensive subject learning so that children will be encouraged to act independently.

In this connection, a list of experts in local communities will be utilized. The list includes experts and scholars with extensive knowledge of insects, wild birds, the cosmos, weather conditions, rivers and geology, automotive engineering, chemicals, and disaster prevention. Through visits, tours, practice sessions, experiences, demonstrations, and lectures, children will be given opportunities to experience the joy of science and technology. At elementary schools, all subjects are taught. Therefore, if an exercise in a Japanese language class is to keep a diary, teachers can let children keep one on insects, plants, animals, etc. As for looking at things from a scientific point of view, teachers can let the children talk, for instance, about an earthquake, and let them compare the objects which fall down easily or collapse with those which do not. In swimming lessons they can look at a picture of a boat to help them think about how they can most effectively move their bodies to increase their speed or improve their buoyancy.

With skills and ideas, teachers should exploit ways to relate other subjects to science and technology education by making them relevant to children's daily lives and by letting them pay attention to the wonders and joys of nature, human beings, and society. It is also possible for teachers to help children discover the principles of nature and allow them to experience the joy of discovery through computer simulation. Children come up with wonderful ideas, and even elementary school children can make ingenious working models out of household materials. It is our hope that, in their daily lives children will freely experiment, observe, learn from practical experience, and experience the joys of discovery and creation.

Selected Bibliography

Chuou Kyoiku Shingikai [Central Council for Education]. 1996. *21 seiki wo Tenbousita Wagakuni no Kyouiku no Aroikata ni tuite* (Dai 1 ji Toushin)" [The Model for Japanese Education in the Perspective of the 21st Century (First Report)]. Ministry of Education, Science, Sports and Culture, Tokyo, Japan, in Japanese.

Chuou Kyoiku Shingikai [Central Council for Education]. 1997. *21 seiki wo Tenbousita Wagakuni no Kyouiku no Aroikata ni tuite* (Dai 2 ji Toushin). [The Model for Japanese Education in the Perspective of the 21st Century (Second Report)]. Ministry of Education, Science, Sports and Culture, Tokyo, Japan, in Japanese.

Government of Japan, Heisei 7nen Houritu dai 130 Gou [1995 Law No.130]. 1995. *Kagaku Gijyutu Kihonhou* [Science and Technology Basic Law], Government of Japan, Tokyo, Japan, in Japanese.

Government of Japan, Kakugi Kettei [Cabinet Decision]. 1996. *Kagaku Gijyutu Kihonnkeikaku* [Science and Technology Basic Plan], Government of Japan, Tokyo, Japan, in Japanese.

Government of Japan, Kakugi Kettei[Cabinet Decision]. 1997. *Keizai Kouzou no Henkaku to Souzou no tameno Koudoukeikaku* [Action Plan for the Reform and Development of the Structure of the Economy], Government of Japan, Tokyo, Japan, in Japanese.

Koudo Jouhou Tuusin Shakai Suisinhonbu [Headquarters for the Promotion of the Advanced Information and Communication Society]. 1995. *Koudo Jouhou Tuusin Shakai ni muketa Kihonhousin* [A Basic Policy for the Promotion of the Advanced Information and Communication Society], Koudo Jouhou Tuusin Shakai Suisinhonbu, Tokyo, Japan, in Japanese.

Marutimedia no Hattenn ni Taiousita Bunkyousisaku no Suisin ni Kansuru Kondankai [Task Force on the Promotion of Educational Policies Adapted to the Development of Multimedia Technology]. 1995. *Marutimedia no Hatten ni Taiou Sita Bunkyousisaku no Suisin ni tuite* [Promotion of Educational Policies adapted to the Development of Multimedia Technology] 1995. Monbushou [Ministry of Education, Science, Sports and Culture], Tokyo, Japan, in Japanese.

Marutimedia wo Katuyousita 21 Seiki no Koutoukyouiku no Arikata ni kansuru Kondan kai [Task Force on Utilizing Multimedia in the Higher Education of the 21st Century]. 1996. *Marutimedia wo Katuyousita 21seiki no Koutoukyouiku no Arikata ni tuite* Monbushou [Ministry of Education, Science, Sports and Culture], Tokyo, Japan, in Japanese.

Monbushou [Ministry of Education, Science, Sports and Culture]. 1995. *Kyouiku Gakujyutu Bunka Supotu Bunnya ni okeru Jouhouka Jissisisin* [Guidlines on the Adaptation of Information Technology in Education, Science, Culture and Sports], Monbushou [Ministry of Education, Science, Sports and Culture], Tokyo, Japan, in Japanese.

Monbushou [Ministry of Education, Science, Sports and Culture]. 1997. Kyouiku Kaikaku Puroguramu [Educational Reform Program], Tokyo, Japan, in Japanese.

Sakamoto, T. 1996. Development of educational technology contributing to educational reform. *Educational Technology Research.* 19: 1-21.

Chapter 18

Links, not Boundaries: An Asian-Pacific Environmental Education Network

Daryl Le Grew, University of Canterbury, Private Bag 4800, Christchurch, New Zealand.

Environmental education is a necessary concomitant of sound environmental management. Creating and disseminating the specialist expertise to identify, model, and solve specific environmental problems in specific locations drives a technical agenda in environmental education and research. Understanding the social and economic systems and the development dynamics within which environmental problems occur drives some aspects of environmental policy development. Ultimately though, expertise and policy cannot have any real impact on environmental futures unless there is an ethical imperative that moderates human intention and action in ways which can guarantee environmental futures.

These are critical factors that each society tends to tackle from within, with occasional reference to international agreements and benchmarks. The complexity and diversity of the international context within which individual nation states must make environmental policy, solve problems and act ethically, however, reaches beyond national sovereignty or the geography of national boundaries. Regional atmospheric and meteorological conditions, ocean currents, geological topographies vary considerably.

Shifts in the big systems of the planet are oblivious to national, ethnic or cultural boundaries. Indeed, the Asian-Pacific Region, while it shares a global context, is influenced by the particular and peculiar conditions of this hemisphere. Recently, business leaders of the APEC (Asia-Pacific Economic Cooperation) countries have noted these differences and called for approaches to environmental management that are regional and that interlink nation states rather than dividing them.

".... the environmental interests and concerns of the peoples of the APEC region are different in kind from the environmental interests and concerns of the industrialized countries of Western Europe. It is important at the global level that these issues are tackled successfully. To do so requires a focus on the particular interests of APEC economies, both in the region and in international fora." (Tasman Institute, 1996: 8).

There are then cogent reasons why the Asian-Pacific Region warrants a distinctive regional approach to environmental management and, concomitantly, to environmental education and research. Clearly, the 1992 Rio Summit underlined global environmental issues and wrought acknowledgment by the assembled heads of government regarding the need for national prioritization and international agreements about environmental benchmarks and targets. What has emerged, though, is a Eurocentric environmental imperative.

"The International environmental agenda is heavily influenced by domestic preoccupations in European and North American countries. Priority issues in international debate such as global warming, ozone depletion and biodiversity reflect these preoccupations." (Tasman Institute, 1996: 5)

These global issues, whilst of seminal importance, tend to be somewhat more abstract, less tangible and less immediate than the environmental problems and imperatives facing the Asian-Pacific Region. They emanate from rapid economic development and the need to address inequalities that flow from sudden industrialization and urbanization. They tend to impact more on the daily lives of people, thereby producing more pragmatic management needs. It is difficult to consider future targets for global warming when poor air and water quality, and waste disposal problems are at one's doorstep.

In their 1993 report to the World Bank on environmental priorities in Asia, Brandon and Ramankutty refer to:

"urban environmental degradation (air and water quality, noise, waste, etc.); industrial pollution, atmospheric emissions, soil erosion and land degradation; water resource degradation; deforestation; loss of natural habitat." (Tasman Institute, 1996: 162)

The authors relate these problems to:

"market and policy failures; the strain on the resource base posed by huge and growing populations; rapid urbanization and industrialization; a common perception that there is a trade-off between environmental protection and economic growth." (Tasman Institute, 1996: 16)

Most nation states in the Asian-Pacific Region would recognize that these environmental problems and their causes are common across the region. Hopefully, they would also recognize that approaches to sharing environmental information and expertise, monitoring and benchmarking environmental quality standards, all point to the need for more cooperative regional strategies for environmental management.

Cooperative strategies emanate from the realization that national boundaries are irrelevant to many of the Asian-Pacific Region's more pressing environmental problems. Air pollution is a case in point. Recent near-catastrophic forest fires that appear to have begun with inappropriate slash-and-burn land clearance practices in one nation, exacerbated by drought conditions and lack of fire management protocols, led to dangerous and debilitating air pollution settling over much of Southeast Asia (Walters, 1997). This is not a new problem, however. Ocean current changes as far away as the eastern side of the Pacific Ocean induce the El Niño effect which sets in place the drought conditions that promote fire; and this coupled with air movement patterns in the vicinity of Indonesia, Singapore, and Malaysia, regularly combine to spread smoke and particulate from forest fires over wide areas. They produce regional pollution levels that threaten health and disrupt local economies, production systems, and tourism, not to mention daily life in several countries of the region.

Air pollution from newly developed industrial and urban complexes and coalfired power stations not only contribute markedly to greenhouse gas emissions, the Eurocentric imperative; but they also contribute to regional particulate and acid rain fall-out that, again, has no respect for national boundaries and national sovereignty.

"Acid rain is also the classic problem of transboundary pollution. Sulfur dioxide and nitrogen oxide emissions from Northeast China fall on Korea and Japan, while emissions from Southeast China fall on Vietnam and Southeast Asia." (Tasman Institute, 1996: 28)

What is true for air pollution also applies to water pollution, especially when nations share common river systems and when ocean currents move pollutants from one national coastline to another. Land clearance and poor irrigation practices upstream in one country induce damaging sedimentation, salination, and environmental impact to countries downstream, and so on. Clearly, in this context regional cooperation on key aspects of environmental management is crucial. However, the need may be obvious; but the will to cooperate all too often arises only as a response to crisis situations such as the recent fires in Kalimantan, Indonesia. Even in this urgent context cooperation is often short-lived and limited to dealing with single problems.

More comprehensive cooperation in environmental management is a regional imperative. What Barbara Lapani (1994) refers to as "enterprise webs" may provide an organizational model that will allow the kind of cross-national cooperative approaches that are needed:

"The same pattern of regionalisation and localism which are emerging from economic globalisation are emerging at the level of organizational structure. Technological innovation and universal competition for market share have increased the need for organisations to pay close attention to the needs of their customers, not just in terms of product, but more importantly in terms of services. As organisations seek to better manage access to rapidly changing knowledge/technology and to link this to the needs of customers they are referring from hierarchies, through flatter customer-focused enterprises, to enterprise webs. Enterprise webs comprise complex networks and linkages to form nodes of value in time and place. Each point on the web represents a unique combination of skills brought together in response to a particular situation. In a complementary way, networks of independent actors .... are forming virtual corporations -- linked together through information sharing, contractual relationships and research and development alliances. These linkages cross organizational, institutional and national borders." (Lapani, 1994: 12-13)

An enterprise web approach to Asian-Pacific cooperation would engender regional agreements on environmental futures together with national and transnational environmental management protocols. Only by such cooperative agreement and action can the Asian-Pacific environment be made sustainable. And sustainability is the key objective of any reasonable environmental policy.

"Sustainability can only make sense at the level of entire systems and not in terms of particular resources, even depletable ones. The energy sources and use patterns of today are not the same as the energy sources and use patterns of a century ago, and even less stability can be expected over the next century. .... A system which encourages enterprise and innovation, which is open and flexible, is far more likely to be sustainable because of greater adaptability than a rigid or closed system....."

"Sustainable development is not about sacrifice for future generations. Sustainable development is about avoiding wasteful use of resources, the crucial point being that we include within judgments about resources use sensible consideration of the interests of our children, grandchildren and great-grandchildren." (Tasman Institute, 1996: 39)

Cooperative management of the Asian-Pacific environment would allow sensible and achievable approaches to setting environmental standards and benchmarks in an appropriate economic and temporal framework. It would allow sophisticated monitoring and environmental modeling, joint approaches to research and development, and the sharing of environmental expertise, cooperative approaches to pin-pointing and dealing with pollutant sources; and it would bring about the shifts in lifestyle, practices and ideologies necessary for a sustainable future. And cooperation would lead to agreements to provide rapid response and technological assistance across the region in times of environmental emergency. Cooperative environmental strategies would influence policy and decision-making at the national level and inform transnational corporations that the Asian-Pacific Region is not a place where environmental standards can be traded for increased profit. Joint and multinational submissions to international agencies, through the United Nations, the World Bank, the Asia Development Bank and foundations can increase the likelihood of cooperatively funded programs to achieve environmental improvement.

But what of environment and educational research? Just as environmental management is best achieved regionally by cooperative mechanisms, so too environmental education can be fostered by such a networked approach. This is compatible with the contemporary view that networked approaches to higher education represent the major focus for institutions in the next decade.

Traditionally, education across national boundaries has been facilitated by exchange agreements for the privileged few; and by open and distance education provided by those institutions that specialize in servicing students in remote locations. However, as Le Grew and Calvert (1997) point out:
> "... the challenges and changes facing higher education are part of an economic, political and social transformation -- reflected in the global economy and the age of information." (Le Grew & Calvert, 1997: 3)

As markets globalize and converge, more sophisticated human interchange networks form that pay no homage to national and cultural boundaries. Trade and commerce are increasingly electronic; resource management and production are conducted globally and without reference to any one national economy; education, information technology, media and entertainment merge as Internet environments are gleaned by an information-hungry world. As part of this new paradigm, universities are also converging educational methods and learning delivery systems to form virtual university networks and hybridized organizations (Slee, 1990) and to partner strategically with private and commercial players (Scott, 1995). Again, Le Grew and Calvert (1997) argue that:
> "Open and flexible higher education is not just a matter for open universities and those that feature distance education; all universities are faced with the issue of how they will relate to a wider clientele and incorporate new methods." (Le Grew and Calvert, 1997: 3)

Open and global systems of higher education assume profiles of specialization and complementarity which, like Lapanian enterprise webs, allow for curriculum development, teaching and learning services and research services to be combined and shared across networks in new patterns of strategic alliance:
> ".... even broader alliances are necessary as institutions realize that they cannot create the resources needed to compete in globalised information environments. Universities may combine with others in national and international consortia to gain more critical mass of infrastructure and intellectual firepower to compete. Equally, these consortia will carry other non-academic partners — publishers, media companies, information technology providers, administrative service

providers, public relations and marketing agencies. Conjointly, international partnerships of this kind will provide opportunities to package together the best of content and academic and administrative services and deliver flexibly to wherever students and organisations are. Being outside the realm of these various international consortia will be difficult for individual institutions, yet negotiating the way through the maze of local and national legislation, international protocols and agreements will require special skills as yet underdeveloped in the higher education realm." (Le Grew and Calvert, 1997: 4)

Consider the possibility of creating this kind of enterprise web — a multinational network of key universities in environmental education, especially those with an interest in the Asia Pacific Region. The critical mass and firepower referred to above would be considerable; the sharing of courses and research, infrastructure, attitudes and approaches, invaluable. Link this network with other players — international agencies and foundations, professional bodies and transnational corporations, the media etc., and embed the network in an Internet environment, and a powerful concept emerges.

Clearly the Internet and the World Wide Web provide the framework for this concept — one in which many players are already working. The Web not only links networking players to one another but can deliver connections to thousands of databases and the web sites of all those other environmental players already referred to, together with computer mediated communication systems, software libraries and information services. A carefully designed and constructed network working on its own Intranet with a delineated range of working functions and a wide array of internal communications can be envisaged — with an equally well designed and delineated set of external connections into the more frenetic Internet environment.

Consider, for example, an Environmental Education Network (EENET). Broadly, an EENET would be formed by a linking of the web sites of, say, institutions represented at this conference — in the first instance. The intention would be to form an international INTRANET — a prioritized, gated and internally managed set of web sites, to facilitate collaborative development, mutual support, joint working arrangements and data sharing that any good Enterprise Web demands.

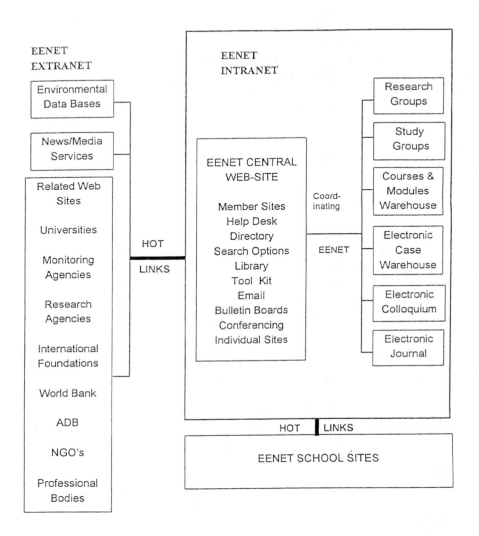

Figure 1. Intranet/Extranet model for environmental education

The EENET Intranet would be divided between a Central Website that links member sites and provides facilitation, assistance and operating mechanisms, and acts as the dynamic valve through which EENET activities, internal and external, could be monitored and managed. As Figure 1 suggests, the EENET Central Site would have a broad range of help desk, directory and search functions; it would access an interlinked EENET library site that combines the environmental collections of individual member's libraries; it would access an EENET Tool Kit of environmental assessment, modeling and benchmarking tools; and, most importantly, it would provide a full array of intra-group electronic communication facilities — email nets, bulletin boards, conferencing and chat facilities; and it would link the web sites and home pages of individuals and sub-groups within the EENET system. EENET would drive interactive and conversational pedagogies replacing didactics with mutual learning, directed access to tools and resources and an ongoing electronic discourse.

The EENET Central Site would also manage a complementary array of specialist functions available to the EENET members, including:
1) Study Groups
2) Research Groups
3) Course Module Warehouse
4) Environment Electronic Case Warehouse
5) Electronic Colloquium
6) Electronic Journal.

Broadly, this set of coordinated functions would ensure that environmental study and research groups, which form among and between international partners have access to an electronic course and module warehouse, which is continuously updated with dynamic multimedia products contributed by the partners and others. As study and research groups work on national and international environmental projects, policy and problem solving, the best outcomes of these would be written as case studies and stored in an EENET Electronic Case Warehouse so that the cases themselves begin to form EENET intellectual property the springing point for new curricula in environmental education, specifically as it relates to the Asian-Pacific Region. In a similar vein the EENET Electronic Colloquium would coordinate the presentation of research projects and papers created by the research and study groups and also act as a forum for the introduction of key consultants, researchers and professionals from outside EENET. This ongoing colloquium would be matched by an EENET Electronic Journal to allow the best of the colloquia presentations to be formalized by a quick response refereeing system into an ongoing intellectual record, for both internal and external dissemination and critical evaluation of knowledge.

The EENET/Intranet would be further complemented by a prioritized set of external connections to create a broader EENET/EXTRANET environment. Hot links with the major international environmental data bases, international news and media services, especially those which produce regular reports on regional development and the environment, and prearranged access to an array of environmentally critical web sites already created by other organizations including, inter alia:
1) Other universities known for environmental research and teaching;
2) Environmental research and monitoring agencies;
3) World Bank and Asia Development Bank;
4) International agencies;
5) Environmental conferences;
6) Professional bodies especially in the environmental disciplines;
7) Foundations and NGOs.

Hot links would allow prioritized access by EENET members through the central web site to this targeted extranet of external resources.

A further extranet connection would hot link the EENET group of universities and others with an international network of schools, especially those that have a strong commitment to environmental education. No doubt teachers in environmental studies and cognate areas would be part of this hot linked environment either for enhancement of their teaching, for pre-service teacher training, or for continuing professional development. Integrating networks of teachers and students into EENET would provide an avenue for increasing awareness of current regional environmental problems, but would also provide opportunities for students and teachers to learn how to cooperate and collaborate, how to share knowledge and experience, and how to use globalization and the information age to facilitate both horizontal and vertical integration of environmental education.

EENET would drive interactive and "conversational" pedagogies that should replace didactics with mutual learning, problem-based and case-oriented curricula. Guided access to internal and external learning resources, teaching linked to current research and a developing library of environmental cases, analysis and design tools and advice would be drawn from the combined resources of an international cross-sectoral learning network, in which self-directed inquiry and extensive discourse would be the essential driving pedagogies. Complementarity, mutuality and cooperation of this kind mark out the future of environmental education and the environment itself.

References

Lapani, B. 1994. *Learning Partnerships: The New Challenge.* Australian Centre for Innovation and International Competitiveness, March 1994.

Le Grew, D. and J.C. Calvert, 1997. Leadership for open and flexible learning in higher education, in *Staff Development Issues in Open and Flexible Education*, Latchem, C. and Lockwood, F. Ed. Routledge, London, UK.

Scott, P. 1995. *The Meanings of Mass Higher Education.* The Society for Research into Higher Education, Buckingham, Open University Press, Bristol, PA, USA.

Slee, P. 1990. Apocalypse now? Where will higher education go in the twenty-first century? in *Industry and Higher Education: Collaboration to improve students' learning and training*, Wright, P.W.G. Ed., The Society for Research into Higher Education and Open University Press, Buckingham, UK, pp 88-91.

Tasman Institute. 1996. Environmental Priorities in Asia and Latin America. A Report to the Monash Group (June 1996).

Walters, P. 1997. Smoke haze kills two as thousands suffer illness. *The Australian.* September 25: 9.

Chapter 19

Biwako-Das: Public Collaboration in Meteorological Observation with a Computer Communication Network

Takashi Toda and Kazuyuki Matsui, Lake Biwa Museum, 1091 Oroshimono-cho, Kusatsu, Shiga 525-0001, Japan and Board of Education, Shiga Prefecture, Kitazakura, Yasu-cho, Yasu-gun, Shiga 520-2321, Japan.

Introduction

Museums have various functions, including collecting, cataloging, research, and education. Exhibitions are part of the museums' education program, the other part being the propagation of knowledge through extension education (see, e.g., Kurata and Yajima, 1997). This wording, however, implies that museums establish a system of knowledge which they then extend to other people (Fig. 1). Some people criticize this wording and suggest that we need an alternative terminology. As a tentative answer to this criticism some museums see themselves as providing a "public service" instead of "education." This implies that the museum is asking the public to collaborate in the establishment of a system of knowledge. In other words, the museum receives information which arises from the public, and acts as a distribution center for it.

This idea is especially important for the Lake Biwa Museum because its main theme is the relationship between people and the lake, and the problem of the environment is an important element in this.

There are many kinds of environmental phenomena that are closely related to human lives. However, local people cannot always relate to them, probably because of their huge scale in terms of time and space, which are much larger than those we usually observe in daily life. In other words, we need a "bird's eye view" of environmental phenomena in order to understand them, while usually we have only a "worm's eye view." Scientists create their "bird's eye view" using special methods, some of the most sophisticated of which use statistics. Here, we first select variables to represent the phenomena to be described (the objects viewed by the bird in Fig. 2a), and then we collect the measured values and process them statistically, creating a mental map similar to that of the bird. One problem which local people experience with this method arises from

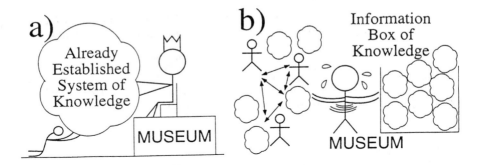

Figure 1. "Extension" and "Public service"; (a) "Extension" in which a museum behaves as if it is an emperor; (b) "Public service" in which a museum behaves as if a servant.

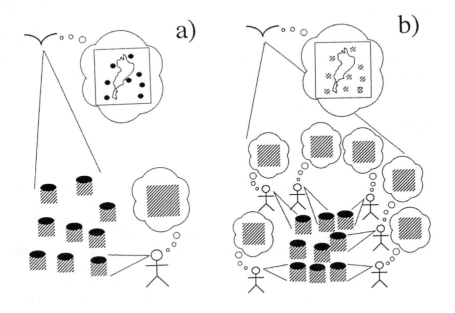

Figure 2. "Bird's eye" and "Worm's eye": views; (a) Usual observation; (b) Observation with public collaboration.

the process of abstraction. The relation of the abstract measurements to the original environmental phenomena is usually very difficult for people with insufficient training in statistics to understand. The results often seem to run counter to our everyday experience (the shaded square in Figure 2a).

A possible solution to this problem is to allow the public to collaborate in the process of abstraction. We call this process "collaboratory observation." First, the local people themselves observe, describe, and discuss the target phenomena based on their everyday experience (the shaded squares in Figure 2b). Next, the secretariat (represented by the bird in Figure 2b) collects these descriptions and makes a summary and statistical analysis. Because the summary and results are closely related to the participants' own experiments in observation, description, and discussion, they can relate to them more easily.

Two Types of Methodology in Collaborative Observation

The Lake Biwa Museum project office started to organize investigations based on collaborative observation in 1992, four years earlier than the establishment of the museum. This early start was based on the project office's idea that the museum's exhibits should be closely related to public collaboration. Because the exhibits provide information to large numbers of people, they are an effective way of telling people about collaborative observation, and to increase the number of participants. The exhibits required some research results, so an early start was necessary.

There are various types of collaborative observation and we can note two examples whose methodologies are at opposite extremes. One is that of biological investigations. The organizers of the investigation (usually members of the museum staff) select each year a target animal or plant species and the method to be used in the investigation. Then they send details of these to the other participants by mail, collect the results by mail, and analyze them to reach a conclusion. Because most of the communication takes place by mail, all classes of people can collaborate in the investigation even if they cannot collaborate in the planning and analysis.

Another type of collaborative observation is that in which the public collaborates in the analysis and discussion of results, and in planning the next steps in the investigation. An important condition necessary to carry out this type of research is a quick means of two-way communication. Participants have to make available their observations for discussion quickly, before other participants forget their own reactions to the phenomena observed. For the same reasons, the organizers should collect the participants' descriptions and summarize them quickly. A computer network provides a useful tool for the following reasons. First, every participant can make available their own information simultaneously to other participants. Second, participants can make available many kinds of information, including text, images, and raw data, and other participants can easily make their own analyses of this information on their own computers. Third, the information can be entered into a database which is a useful resource either for discussion or summarizing the data.

Biwako-Das is an experimental project in collaborative observation based on a computer network dealing with the meteorology of the Lake Biwa region.

The Pre-history of Biwako-Das

Biwako-Das arose out of two previous projects, Yuki-Das and Hotaru-Das, which were started in the summer of 1989 and continued to the winter of 1991-1992 (Kada and Oonishi, 1991). These were planned as collaborative observation of two dramatic and familiar environmental phenomena, the appearance of "hotaru" (fireflies) in early summer and "yuki" (snowfall) in winter. The pairing of "hotaru" and "yuki" is traditional in Chinese culture, and this was another reason for selecting these two phenomena. Because the AMeDAS (Automatic Meteorological Data Acquisition System) of the JMA (Japan Meteorological Agency) had become well known at about that time, these two observation projects were named "Hotaru-Das" and "Yuki-Das," respectively.

In the discussions which formed part of Yuki-Das, the participants were interested in the relation of wind to the snowfall as well as fishermen's folklore about this relationship. Because both traveling on the lake and fishing are strongly affected by weather conditions, fishermen must try to predict them in order to make their own voyages safe and plan their fishing.

In the folklore each wind pattern has a name. One of those names which has a close relation to snowfall is "Hiarashi," a light wind blowing from the south or southwest along the long axis of Lake Biwa. They say that a heavy snowfall follows a Hiarashi which blows continuously for a few days. The wind pattern occurs when a strong low pressure is located to the northwest and the arrival of the low pressure causes heavy snow. The word "Hiarashi" can be interpreted as a combination of "Hi" (sunshine) and "Arashi" (breaker).

Through the discussion of this folklore, the participants in Yuki-Das recognized the importance of the wind for the meteorological conditions and began to want to become involved in their own wind observation.

The Progress Towards the Biwako-Das System

In the planning of the Biwako-Das wind observation system, our aims were as follows. First, because we were interested in the local meteorological conditions in the Lake Biwa area, it is important to recognize different wind patterns at different points around the lake. Second, because we wanted to base our discussions on both the observed wind data and people's experiences of daily life, a real-time system of data acquisition was necessary, i.e., we needed to collect observations at every point every few days. Third, in order to expand public collaboration, we wanted to display summaries of the data publicly, for instance as an exhibit at the Lake Biwa Museum, in order to arouse people's interest and to recruit new participants.

The data from the Japan Meteorological Agency were insufficient for our requirements because they are collected for general use, especially for weather forecasting and related mesoscale meteorology, therefore, have insufficient observation points for our purposes. In addition, there are severe technical and bureaucratic problems in real time data acquisition for any purposes except weather forecasts.

Most other wind data are collected in connection with disaster warnings are not recorded during normal calm weather. There are some observations collected by environmental agencies but the technical and bureaucratic problems in getting access to them were too severe when we started Biwako-Das.

We, therefore, decided to observe the wind by ourselves. As a test case, a wind observation system was set up at the private house of a participant in 1992. In the Yuki-Das project, it was not necessary to construct an automatic observation system because we needed only daily snowfall data together with daily maximum and minimum temperatures. In Biwako-Das, in contrast, it was important to record hour-to-hour changes in the wind, and so automatic data acquisition became necessary. We, therefore, tried to set up an observation and data recording system, together with a communications system through which we could exchange and discuss the information recorded.

This trial stage continued for about two years and then we started to set up other observation stations. In 1996, a Biwako-Das system

258 Integrated Environmental Management

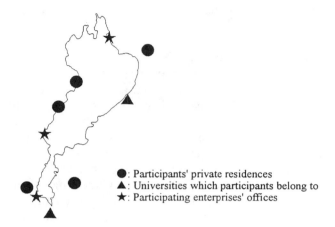

Figure 3. Observational stations for Biwako-Das

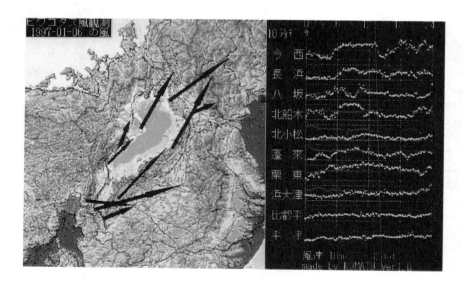

Figure 4. An example of Biwako-Das wind data, January 6, 1997: a typical example of the "Hiarashi" wind pattern.

was set up with ten stations (Fig. 3). Five of the ten stations are set up in the private houses of participants. Two others are set up at the universities to which participants belong, and three are at the offices of participating enterprises.

Results to Date and Possible Future Developments

Because the full Biwako-Das system just started in the last year, we are still interested in the way in which the whole system can be maintained. Fortunately, we still have not experienced severe problems and the system is running well.

We are also collecting information on typical wind patterns, which are useful in generating public interest. As an example, data on a typical "Hiarashi" are shown in Figure 4.

The Biwako-Das system has started to provide other information on regional meteorological processes, including the following:
1) the appearance of land and lake breezes, their relation to weather conditions, their difference from point to point, etc;
2) the structure of the meteorology of the region, especially the distribution of wind, as a function of mesoscale meteorological conditions;
3) the relation of the flow of water in the lake to the change of wind pattern on the lake;
4) the effects of the wind system on the distribution of water quality, biomass, and the ecosystem;
5) the explanation and confirmation of folklore relating to meteorological conditions; and
6) the relation of everyday experience to the systematic meteorological data.

We still have not decided the next goal of Biwako-Das, but we can enumerate several possibilities as follows based on our experience to date:
1) Establishment of ways of systematizing personal meteorological observations, and of exchanging observational data through the computer communication system, by sorting, analyzing, generalizing and theorizing the problems which have occurred in building up the Biwako-Das system;
2) Development, generalization, and theorization of methods of displaying data;
3) Examination of ways of developing the observation network;
4) Examination of additional stations for other types of meteorological phenomena; and

5) Examination of the possibilities of making use of observations from other organization, e.g., from the environmental agencies, and how to overcome the bureaucratic obstacles to real time data acquisition.

Problems of Biwako-Das

As is mentioned above, Biwako-Das was planned as collaborative observation with in-depth participation, that is to say in which the public can collaborate in all stages, including planning, observation, and analysis. In order to achieve this depth, participants should ideally have skills such as the ability to use a computer communication network and the ability to set up and maintain the meteorological equipment. This makes it difficult to recruit new members. For the same reason, it also depends on the energy and time of existing participants. In fact, the analysis of the results is frequently halted when leading members are busy with their jobs. As a result, the progress of Biwako-Das as a research project is extremely slow compared to the usual academic investigations. These two problems are severe, but they are an inevitable part of investigations like Biwako-Das. We predicted that we would experience them, but as yet we have found no effective solutions.

Acknowledgments

The authors thank all the other Biwako-Das participants for their collaboration in the planning, development, maintenance, and discussion processes, especially to the planning work by Dr. H. Nishinosono, Dr. Y. Oonishi, and Dr. Y. Kada. We are also grateful to Mr. H. Takada who discussed and advised us on meteorology, and to Mr. K. Kawai and Mr. K. Ookawa for setting up and maintaining the earlier observational stations.

References

Kada, Y. and Y. Oonishi. 1991. Exploring the materials in our neighborhood environment—the purpose and process of the "Firefly-Snow Strategy." *Japan Society for Science Education Report*, 6(1): 45-50 (in Japanese).

Kurata, K. and K. Yajima. 1997. *Shimpen Hakubutsukangaku. [Museology* (New Edition)]. Tokyo-do Shuppan, Tokyo, Japan, in Japanese.

Chapter 20

Establishing a Course in the Practical Use of Media Tools at Shiga University — Introduction of a Flexible Learning Program

Hajime Hirai, Faculty of Education, Shiga University, 2-5-1 Hiratsu, Otsu, Shiga Prefecture, 520-0862, Japan.

Computer Illiteracy

There is no doubt that the significance of advanced technologies in digital information and communication, including the Internet, will increase among university students, not only for research and educational purposes, but also in everyday life activities such as hobbies and job searches. Institutes of higher learning have predicted the coming of the digital network society and are planning a better environment for the use of the Internet. It may not be too exaggerated to say that the Internet provides the potential for universities to survive and advance in the coming 21st Century.

At the Otsu Campus of Shiga University, where the Faculty of Education is located, enormous efforts are made to provide an infrastructure for the local area network (LAN) and Internet systems. Most of offices and some classrooms are now fully connected by a LAN system. A total of more than 100 IBM compatible PCs and about 30 Macintosh computers are available exclusively for students at the Otsu campus.

The Information Processing Center issues identification numbers for students to communicate by email and to open their own home pages using the student server. The center also has started providing services for connections from off-campus sites. Two technicians and about ten faculty members, who are members of the center's management committee, manage and maintain computers and the network system on daily base. We consider that the digital network systems on the Otsu campus are developing quite effectively and functioning relatively well, in spite of budgetary and manpower shortages.

It is unfortunate, however, to have to admit that not as many students are taking advantages of these relatively privileged circumstances as we might have expected. If this is correct, why don't they use the facilities? Aren't they good at science and mathematics? Aren't they interested in computers and the Internet? Or are they too busy with extra-curricular activities and part time jobs? This was a puzzle for me until I finally found a clue from conversations with colleagues and students. What I found was that they are neither poor at science nor indifferent to computers. They are aware that computers and digital networks will become important and critical for their future careers, and they are eager to learn how to use them. The biggest problem for them is, however, finding the opportunity to learn to use them in a systematic way, as this was not being provided by the university.

Most students realize the necessity of having a good command of "media tools" (i.e. the knowledge of how to use computers and the network system) when they start preparing for their graduate theses in their third or fourth year of university education. Then, they start learning about them more or less by themselves and with no organizational support from the university. This is neither a wise nor effective way to master such highly complicated and sophisticated technologies. It is no exaggeration to say that we may have been neglecting our educational mission by not providing students with proper services.

Practical Use of Media Tools

In April, 1997, a new attempt started on Otsu campus. The Faculty of Education started a course called "The Practical Use of Media Tools." The class is primarily offered to the first year students who have registered in the teacher training program on Otsu campus.

The main purpose of the course is to provide students with opportunities to learn how to use computers and computer networks in a systematic way. We have adopted so called "team teaching" methods for this class. A total of 24 faculty members, at least one from each department, participated in the steering committee. Seven members of the committee were appointed as members of the "core group," whose roles include setting an overall timetable, selecting possible topics to learn, and preparing the syllabus. They are also responsible for organizing guidance sessions and evaluating grades. Members of the steering committee are expected to be involved in the course by assisting students in their departments when the students need help.

The class may be unconventional in terms of its curriculum and teaching methods, which we call "flexible teaching and learning." For instance, instructors and students do not meet each other at a fixed

Table 1. Curriculum (Basic A: Compulsory)

Unit No.	Topic	Description	Points
A1	Guidance 1	Overview of computer and network system. Introduction to Windows	2
A2	Guidance 2	How to start, manipulate and shut down Windows	4
A3	Internet: Basics	How to use the browser. Searching for sites and collecting information.	4
A4	Word Processing 1	Typing, printing and file saving.	4
A5	Electronic Mail	Setting up email preferences. Sending/ receiving mail.	4
A6	Word Processing 2	How to write a report. Page layout and editing.	4
A7	Term Paper	Submitting a term paper.	2

time and place. In "Basic A," the students are not required to attend any classroom activities except for the first two guidance classes. They either bring assignments in print or send them by email to one of the core group members in charge of the unit. In "Basic B," they may select any topics they wish in any order. It depends on the instructors in each unit whether they request students to attend the class or not, whether they set a certain timetable to complete the unit, and what kind of media they choose when assignments are submitted.

The class consists of a total of 24 units, of which seven are labeled as "Basic A" while the other 17 are "Basic B." The former is compulsory and the latter is elective (see Tables 1 and 2). Units in "Basic A" cover very basic skills such as word processing, sending and receiving electronic mail, and browsing the World Wide Web (WWW). Students are expected to take "Basic A" as early as possible even though no particular date is set for the completion of each unit. Once they have completed "Basic A," they will move on to "Basic B."

Table 2. Curriculum (Basic B: Elective)

Unit No.	Topic	Description	Points
B1	Spreadsheet: basics	Introduction to data analysis using *Excel*	4
B2	Macintosh: Basics	Sending and receiving email and files using *Eudora*	4
B3	Databases	Designing an address book using *File Maker Pro*	4
B4	Graphics	Introduction to graphic design.	4
B5	3D graphics	Introduction to 3D graphic design.	4
B6	Sending information by Internet	How to make a home page,	4
B7	UNIX basics	UNIX and its work station.	4
B8	Internet discussion	Participating in "Discussion Groups."	4
B9	Graphic processing	Graphic data processing using *Photoshop*.	4
B10	Computing	Computing by and for the handicapped.	4
B11	Calculation	How to use a functional calculator.	4
B12	Data processing	Practical use of data processing in health science.	4
B13	Desk Top Publishing	Designing posters and newsletters using *Microsoft*	4
B14	BASIC	Programming using BASIC.	4
B15	Spreadsheet: statistics	Statistical analysis and graphs using *Excel*.	4
B16	English word processing	Writing and editing using an English word processor.	4
B17	Spreadsheet: applied	Data analysis in life science using *Excel*	4

Table 3. Grading

Grades	Points earned
Excellent	30 or more
Good	26-29
Pass	22-25
Fail	21 or less

In "Basic B," students may wish to take any unit listed in the guidebook, based on their own interests and requirements. The steering committee made an effort to include as many kind of topics as possible, because students are expected to come from fifteen different subject majors.

The committee was successful in organizing 17 units in "Basic B", covering areas such as English word processing, graphic design, producing a home page and statistical analysis. Based on their accumulated points, they automatically receive the appropriate grade (see Table 3). Students will normally get four points when they finish one unit. Those receiving a total of 30 points or more will automatically get an A grade. Those from 29 to 26 points will get a B grade while those from 25 to 22 will get a C. Those who fail to get 22 points or more will not pass the course.

Our message to students is quite simple and clear. That is, "It is up to your interests and efforts how many points you get and what you learn from the class. We will support you. But it is you who will do it by yourself." Of 208 students, 206 registered for the class.

Problems to Be Solved

Some traditional and somewhat conservative faculty and staff members thought the course was too unconventional and radical to fit into the curriculum. They were bewildered to find that the main purpose of the class is neither to provide knowledge of a certain academic field nor theory and methodology, but tools, methods and skills for communication. They thought that these were peripheral and did not fit in with university education. Officers at the Office of

Student Affairs were anxious about the course, too, because students were not required to attend the class at a certain time and place. They were worried how the accumulative points system works.

What some faculty and staff members were worrying about was understandable to a certain extent, because the concepts on which this course was based were totally new in this university. But we attempted to reassure them by saying that what we were trying to do was not eccentric, but a reflection of the demands of society.

In fact, however, there were many problems to be solved before the course commenced. First, the number of students who will register for the course, about 200 every year, is expected to be quite large. There is no classroom on campus where such a large number of students are able to use computers at the same time. Second, it is almost impossible, from a curriculum management viewpoint, to put such a large class into a particular time slot. There is no single academic member of staff whose responsibilities include teaching this kind of media tools course to such a large number of students.

After a long discussion among the core group members, we concluded that the class should be managed flexibly in terms of not only the content but also teaching and learning methods. We decided the class should be organized by a team of faculty members, in a team teaching format, and that the individual members would be responsible only for the units they were in charge of. We regard the members not as teachers but as coordinators, setting assignments, giving advice to students on solving problems, and maintaining the software and hardware necessary for the assignments.

Since the class started in April 1997, we have found several problems, some of which must be solved in the shorter term, and other in the longer term. The biggest difficulty has come from the absence of staff to supervise and coordinate the class overall. Even though each member of the group should be, and is, able to be responsible for each unit they are in charge of, the class cannot be managed overall without a coordinator on a full-time basis.

The next problems are those of hardware and software. We have faced more technical trouble with computers than we had anticipated, some of which we had never experienced before. Having adequate manpower to maintain the technical aspects of the computers and networks is crucial to meeting the demands of users. The financial backup is important, too. Technologies advance so fast that many facilities very soon become out of date. New types of software are being developed and released on an almost daily base. It is almost impossible to keep up with such information and to purchase and download the software, from the viewpoints of both finance and manpower, given that the class is maintained by volunteers without a full time coordinator.

Developing and refining the curriculum and course guide book will be another huge task as we try to improve the quality of the class. There has always been a debate about what is necessary and desirable in the media tool class. The demands of students are wide and varied, because they come from fifteen different departments. Rapid advances in technology and the market in media tools also make it more difficult to decide what software to use. There are many barriers to clear in order to adopt the new technologies, as I mentioned above.

Revising the course guide book every year could be a huge extra task for members of the steering committee, because they already have teaching and research obligations in their own fields. In a relatively small institution like Shiga University, it is almost impossible to hire teaching staff exclusively for this subject.

The motivation among the students registered in the class varies, from those with definite expectations and clear motivation, to those who have registered in the class only because it is one of the required courses for graduation. Some of the course's characteristics, for instance the lack of fixed deadlines for assignments and the lack of face-to-face teaching methods, may have worked negatively, particularly for those students in the latter category. On the other hand, for those in the former category, the demands made by most of the units may not be high enough. I think that for those highly motivated and capable students we should discuss what we can do to meet their demands within a broader framework, that is at the levels of the university and/or the faculty, and not within one single class.

Future Developments: Towards Flexible Learning

Probably the most important and urgent agenda to discuss is: what comes next after learning these tools? What we are expecting in this class is that students will acquire the basic skills and knowledge. Then, the possible next step for us is to provide students with opportunities to apply these skills and knowledge to their interests. In other words, we need to create a setting in which students can maximize what they have learned in this course.

There is no question that we should pursue the possibility of the advanced use of media tools in traditional research and educational settings. Students will use computers and the network system more often when they are engaged in writing term papers and conducting research for their graduation theses. But this is not enough. The environment surrounding institutions of higher learning has changed drastically due to technological advancements in information science. Technologically mediated forms of education facilitate greater flexibility in the time and/or space of teaching and learning and in the provision of resource based forms of teaching suitable for different contexts and student groups. "Flexible learning and teaching" may be the key concepts for educational opportunities in the near future.

The term, "flexible learning and teaching" may be defined differently from by different people. However, the vision is fundamentally that of making the university's course offerings more accessible to a more diverse range of student groups studying at the undergraduate, postgraduate and advanced professional level in a broader range of educational settings. These may include not only the campus, but also cross-campus, off-campus, workplace, and home settings. In order to make flexible learning and teaching more comfortable and effective, various types of technology, such as audio visual technologies, computer assisted learning, and interactive multimedia learning resources are becoming primary tools.

Advances in technologies and the concept of flexible learning and teaching may change what has been negative in educational settings to something positive in an age when boundaries in time and space are either disappearing or meaning less and less. For instance, historically many people at Shiga University have considered that having two campuses 60 kilometers apart causes more inconvenience than convenience, as if these campuses are different institutions. It is extremely difficult for students on one campus to enroll in classes at the other campus. Faculty members seldom teach at both campus unless there is a special reason to do so.

As far as postgraduate programs in the faculty of education are concerned, those who teach at local elementary and secondary schools are allowed to study on campus only for the first year of graduate school. They are expected to go back to their work place when they enter the second year, after which they have very limited opportunities to study on campus. This situation puts both students and their academic advisers in an extremely difficult situation when they are preparing their master's thesis.

The flexible learning and teaching methods may help solve these problems. Knowledge and skills with computers and computer networks will definitely be one of the key factors in promoting the new type of learning and teaching environment. We hope that this course will serve as the first step toward more flexible learning and teaching for all students no matter where they study, no matter what their major subject is, and no matter at what levels they are studying.

Selected Bibliography

Deakin Centre for Academic Development. 1996. *Flexible Teaching and Learning at Deakin University*. Deakin University, Geelong, Victoria, Australia.

Young, J. R. 1997. Rethinking the role of the professor in a age of high-tech tools. *The Chronicle of Higher Education*, October 3, A26-A28.

Akagi, S. 1996. *Internet Shakairon* [The Internet Society]. Iwanami Shoten, Tokyo, Japan.

Nakamura, E. and H. Komon. 1995. *Multi-media ga Kyouiku wo Kaeru* [Multi-media changes in education]. Nikkan Kogyo Shinbunsha, Tokyo, Japan.

Chapter 21

What Education and Information Systems Can Do to Help Solve Environmental Problems: A Summary

Frank M. D'Itri, Institute of Water Research and Department of Fisheries and Wildlife, Michigan State University, East Lansing, Michigan 48823-5243, USA.

In this symposium the battle cry to "think globally and act locally" offered a fruitful framework in which to interrogate the situation in the Asian-Pacific region. The participants considered not only environmental and educational problems but also ways to apply education and information systems to resolve them.

In the twenty years since the International Environmental Education Workshop was held in Belgrade in 1975, substantial efforts have been undertaken to make the world's population aware of and concerned about the environment and its associated problems. However, since the Belgrade Charter and the Tbilisi Intergovernmental Conference Report both listed environmental education objectives in 1977, criticism has also been substantial. Environmental education has been charged with lacking direction and not helping students develop the skills to solve environmental problems (Hungerford and Volk, 1990). The presentations at this conference indicate that major steps are being taken toward resolving some of the issues as nations, groups, and individuals begin to work together in a global framework to enrich one another's understanding of new developments.

Concern for environmental degradation and research to improve water quality were already well underway before new global policy decisions such as Agenda 21 and Agenda 21+5 emerged from the 1992 Rio de Janeiro conference. This global meeting sounded a call for all nations to examine their environments more carefully and exchange ideas and technical support to raise awareness of the problems and seek solutions. The issue is complex because developing nations experience internal pressure to catch up with industrialized nations in economic achievements as well as external pressure to accommodate economic globalization. As the public raises expectations for a higher standard of living, awareness of environmental issues may not increase at an equivalent pace. Also, although countermeasures may be initiated to combat environmental degradation and institutional measures such as new laws may take effect, the need for financial resources continues to inhibit progress.

Efforts to cope with problems created by industrialization and the depletion of natural resources have varied locally, regionally, nationally, and internationally. At this conference these were identified in two major areas: citizens' concerns and education. Participants noted how NGOs and other citizen organizations have played an important part in raising awareness and attempting to halt further environmental degradation. In some cases they have done so directly such as by trying to save forests from destruction. In others, protests and other pressures have been exerted on governments to set limits on pollution and to institute cleanups. In this regard, a lack of finances has been a common problem, and sustainable development a common objective.

The concept of sustainable development was launched at the UN Conference on the Human Environment in Stockholm in 1972. The term, "sustainable development," was introduced in the World Conservation Strategy in 1980. Under this concept the internal resilience of a system or a country must be strengthened in a proper balance between humanity and the environment. These issues were brought together at the World Commission on Environment and Development in 1987.

Sustainable development depends on proper support for the social, natural, and man made environments. For example, some economies are not sustainable because the natural capital is depreciating at a faster pace than savings. This issue has to be examined on a global scale as other countries that import raw materials for manufacture are able to maintain a savings balance. Instead, a more equal partnership is needed among nations to maintain sustainable development and alleviate poverty. This has to include appropriate education and proportionate participation internationally to improve the quality of life for all.

While the participants agreed that problems cross boundaries and can be more readily resolved with international cooperation, some differences necessitate that many issues be addressed locally. For example, where autonomous local civic structures are developing, citizens require education in how to function within them. In other cases organizations have studied environmental problems for a relatively long time, but their nature and content have changed with rising awareness of different kinds of issues. Many Asian countries have experienced a major shift recently in both pollution problems and in the level of awareness and public concern.

Consequently, much attention has been focused on air pollution, acid rain, contamination of water/soil by artificial toxins and other substances that originate from secondary industrial production. However, a more immediate concern might well be the devastation of seminatural ecosystems such as agricultural/pastoral lands in developing regions. For a long time, human activities and natural processes maintained a balance that allowed sustainable production. However, the invasion of the market economy and over-harvesting have resulted in the collapse of such balanced systems and paved the way for a shortage of food and water resources in the near future. Consequently, the restructuring of primary industry and the restoration of sustainability have been proposed as priorities.

The need for local, regional, and national planning was emphasized as research on environmental degradation was described for Lake Biwa in Japan, the Indus Valley in Pakistan, and the nation of Malaysia. At Lake Biwa, lake management strategies have evolved as issues such as the discharge and control of atmospheric pollutants, solid wastes, and hazardous and toxic chemicals were recognized, as well as the need to manage liquid wastes. Other transforming regions in Asia have parallel problems and all can benefit from exchanges of information. In Malaysia, increasing industrial development has led to a decline in natural resources and a deterioration of the environment that prompted legislation to initiate more effective use and protection. In a series of five-year plans implemented since 1957, Malaysia has replaced rubber and tin production with manufacturing as the dominant economic factor. However, the detrimental effects have raised the need to redress the depletion of resources and deterioration of the environment, notably the rainforest, air, and water. The effort has shifted from "curative" to "preventive" measures; but more public awareness is needed.

In Pakistan, drainage from the Indus Valley to the Arabian Sea has been disrupted as deforestation on mountains has had an impact on rivers and coastal areas. Soil carried down the rivers denudes the mountains, clogs dams, and causes flooding of the plains. A positive effect is that agricultural soil is replenished, but inadequate irrigation may permit salt to accumulate. Despite the complexity of these problems, an even greater threat may be posed by global warming, which changes the climate and may result in flooding of coastal areas.

Recognition of environmental problems emerged with attempts to industrialize in the Republic of Korea. These were augmented by the government's cheap energy policy and added to public distrust of the drinking water supply. Democratization in 1987 led to citizens' efforts to investigate pollution damage, monitor the environment, and engage in other activist causes. Awareness raised initially with local environmental problems has extended to regional, national, and global issues.

Participants in this conference observed that research can identify the problems and legislation may be passed to slow down degradation and restore the environment. However, whether it be lake waters or deforestation, support for paying the cost of restoration requires an informed citizenry. Consequently, environmental education has become a major priority often led by environmental NGOs and other citizen organizations. With the ever burgeoning information technology, environmental networks and distance education are constantly being updated and extended. Therefore, new methods of teaching, learning, and improving educational materials and methods were presented. These encouraged participants to consider new ways to promote environmental education both formally and informally.

Methods of teaching and learning through distance education have changed in response to new technology: first telephones and radios, then television and computers. But the new research and educational methods still leave many problems unresolved. More and better materials are needed to educate the public on the complexity of environmental issues and how much remains to be done. A renewed effort was initiated in 1989 when the International Lake Environment Committee (ILEC) implemented a seven year pilot project in primary and secondary schools in six countries with widely different natural, cultural, and socioeconomic backgrounds. The emphasis was on local problems. In one instance, a nearby water source was studied and this initiated recognition among citizens of the function of natural ecosystems and the human impacts on them. This very successful approach could be a model for future environmental education.

In Japan, awareness of pollution problems was generated in the mid 1950s. Citizens of Shiga Prefecture then started education projects on public hazards and how to protect nature. Because the core issue of environmental education was recognized and adopted, the Prefectural Board of Education has published handbooks and a sub-textbook to help teachers. Model schools supply materials that teachers integrate into their curricula. One popular example is The Lake Biwa Floating School on the ship "Uminoko." Students actually go out on the lake and conduct field studies. They also work with local citizens to preserve the environment. In this manner, participatory learning is a way to nurture problem solving abilities and expressiveness.

Other major changes are also being recommended to enable the Japanese educational system to allow greater flexibility and encourage a zest for living among school children. As programs have been implemented in all public, national, and private schools in Shiga Prefecture, teaching methods and techniques have improved. In-service training is being provided as well as study conferences for teachers in charge of environmental education. Guidelines to incorporate global environmental problems were offered in 1991, but better teaching materials and curricula are needed as well as more emphasis on lifelong education.

Centers like the one at Shiga University help re-educate teachers and other citizens and develop new teaching materials. The incorporation of local daily life materials along with direct experiments, observations, and research nurtures an inquiring attitude. The Center plans to create a database of teaching materials and programs to offer schools. Local environmental education programs are also being extended among Japanese adults. Citizens have participated in wind measuring experiments, and the data have been displayed at the Lake Biwa Museum. This direct citizen participation in weather data collection has a practical application for fishermen. The project also permits the assertion of individual responsibility and self actualization on the part of citizens, as well the achievement of local recognition and generation of civic pride.

Similarly, an introductory course on how to use computers has aided students at Shiga University and opened possibilities to facilitate greater flexibility in teaching and learning. Media programs facilitate learning on both Shiga University campuses, and course offerings are accessible to a more diverse range of students in varied settings. For instance, secondary school teachers may be able to take graduate courses off campus. Thus, while this new course was originally designed to facilitate learning basic computer skills, it is opening up opportunities that were not apparent initially. Among the long range prospects are improvements in the level of teaching in elementary and secondary schools as well as in the university.

New textbooks and teacher training programs are also being developed to implement the environmental education values established under the 1992 National Environmental Education Action Plan in the Philippines. More and better teaching materials are needed although environmental education has been underway for decades. Malaysia also has in place a substantial environmental program from primary school to university. Teacher training programs have been designed to help integrate environmental issues in school curricula. The essential factor is for teachers to develop a holistic understanding so students can be given practical applications. For example, a community based "River Watch" project monitors the physical, chemical, and biological quality of water in a river near the university. Besides regular courses, NGOs and the Malaysian Nature Society cooperate with the government and private sector in education field centers and other projects.

As in Malaysia, the Philippines, and Japan, Indonesia has an environmental education program underway with an emphasis on sustainable development. Teachers in Thailand also have attempted to integrate environmental facts and concepts throughout the curriculum. Environmental studies have been offered as part of the required general science syllabus and in special classes. Environmental education specialists favor the more integrated approach wherein interdisciplinary planning teams devise problem based units or courses. However, as in Shiga Prefecture, a major problem is to find time within the teachers' already full schedules.

Participants agreed that environmental education requires new techniques to identify pollution problems and countermeasures to solve them. One problem is to differentiate between problems that have to be solved by specialists and those that may be resolved by mutual consent or simple decision making among stake holders. The big challenge now is to develop the flexibility of attitude, motivation, and techniques to identify appropriate aspects of environmental education. These can be addressed in various circumstances and ways devised not only to disseminate information but also to enlist cooperation and support.

As children need room to learn to dream and be creative, so do teachers, researchers, and others. One current tool with great possibilities is the Internet. Applications can be designed specifically for the Asian-Pacific Region. Further cooperation is required to solve international problems related to atmospheric, oceanic, and geological conditions that cross national, ethnic, and cultural boundaries. So-called "enterprise webs" may facilitate this cooperation because they cross organizational, institutional, and national boundaries to share information, establish contractual relationships, and form research and development alliances. Communication networks with new patterns of strategic alliance can enable universities to form consortia among themselves and with non-academic partners, to package and present the best academic resources.

The World Wide Web can provide the framework to connect participants. An environmental education network can link web sites and facilitate collaborative development, mutual support, joint working arrangements, and data sharing. Such a prioritized linkage of web sites could direct participants to the best results of national and international projects, generate study groups, and even publish journals and warehouse the best information, a library on the net as it were. Hot links through the central web site would allow prioritized access to major international environmental databases and other significant resources. This kind of network has the potential not only to allow greater collaboration but also to facilitate joint efforts to make sense out of the constantly proliferating information sources on the web.

Numerous examples were given in this conference of new kinds of cooperation among citizens, scientists, and government officials such as more flexible education that can lead to mental rehabilitation, what was called "a zest for living," as well as improvements in the environment. The participants expressed a renewed enthusiasm to cooperate and encourage others to examine issues from one another's perspectives, to understand the desire for economic equality, and to improve the standard of living of human beings overall. At the same time they developed a better understanding of the tradeoffs in environmental degradation and the need to formulate plans for sustainable development. This conference represented a further step in international cooperation in the Asian-Pacific Region through better understanding of one another's problems and efforts to solve them as information is exchanged through environmental education.

Reference

Hungerford, H.R. and T.L. Volk. 1990. Changing Learner Behavior Through Environmental Education. *Journal of Environmental Education*, 21(3): 8-21.

INDEX

-A-

academic journals 216
ACHEE (Asian Committee of Higher Environmental Education) 207
acid rain 12, 32, 132, 143, 189, 201, 235, 243, 273
acidity 5, 103, 141
Aegiceras corniculatum 67
aerosols 8
affluence 95
Afghanistan 54
Agenda 21   24, 193, 271
aging society 230
agriculture 31, 45, 94, 96, 137, 187; and land 8, 11-12, 23, 31, 50, 54, 58, 61, 99, 114, 184
air 2, 4, 7-8, 12, 23, 26-27, 32, 36, 78-79, 90, 104,107, 109, 112, 116-17, 123, 132, 137, 151, 189, 205, 242-243, 273
airports 17
algae 139, 185
allonomy 206
Almaden Research Center 218
Alta Vista 212, 214
aluminum 221
AMeDAS 256
American Automobile Manufacturers Association 219, 221
American Chemical Society 216
animals 54, 97, 190, 237; see also *fauna*
Anmyun Island 81
Aogashima Elementary School 235
Aoko 153
APEC (Asia-Pacific Cooperation) 241
aquaculture 9, 31, 101, 102, 187
aquatic environments 131, 137
Arabian Sea 62, 65, 68, 273
Aral Sea 138
archaeology 71
Asian-Pacific region 224, 227, 241-45, 271, 276-77
Asian Development Bank 169, 245, 250
Association of Literature and Environment 207

audio files 214
automotive industry 221
Avicennia marina 67
awareness 82, 90

-B-

bactericides 218
Baikal, Lake 27
Bakun Dam 31
balance of nature 174
Balochistan 64
bandwidth 217
Bang Na 101
Bangkok 67, 99,101, 103-12, 117, 119, 123, 126
Bangladesh 51
Bansong-dong 81
Bar Valley 55
barrages 58-59
Basic Law for Environmental Pollution Control 144
Beijing 11, 15, 17
Belgrade Charter 143, 171, 271
biochemical oxygen demand (BOD) 102-05
biodiversity 4-5, 8-9, 24, 27, 31, 49, 58, 70, 96-97, 117, 137
biomass 259
bird watching 124
bird's eye view 253-54
birds 54
Biwa, Lake (Biwako) 19-28, 138-39, 150-51, 153, 156, 203, 224, 273-75
Biwako-Das 256-60
Brahmaputra 51, 57, 68
breezes 259
browser 213
Brundtland Commission 190
bubble economy 17
Buddhism 182, 224
bullying 229
Bureau of Environmental Protection 9
bureaucracy 17, 114, 119-20
businesses 86, 226; see also *private sector*

-C-

California State University, Sacramento (CSUS) 220
Cambridgesoft 219
campaigns 173
canal systems 58, 60-61, 65, 103
car emissions 8
carbon dioxide 12, 69, 119, 201
carbon monoxide 8, 108
carrying capacity 206
censorship 4
Center for Environmental Education and Lake Science 150, 152
Central Council for Education 157, 160, 229, 230
Central Plain (Thailand) 101, 181
Central Region 96, 100
Centre for the Study of Environmental Change 217
Ceriops tagal 67
Chacherngsao Province 184
Chao Phraya River 104, 117, 123
ChemConnect 220
ChemCenter 216
Chemical Manufacturers Association 219
chemicals 7, 9, 22, 136, 221; spills 78, 82; hazards 204; see also *fertilizers, wastes*
ChemNet 221
ChemTrade 221
Chernobyl 10
Chiang Ching-kuo 9
Chiang Jiang Dam 137-38
Chiang Kai-shek 9
Chiang Mai 99, 151, 166, 183
Chiang Mai University 165
China 1, 4, 6, 7, 11-15, 51, 79, 93, 137, 243
Chitral 55
chlorofluorocarbons 8, 37, 68-69
Chon Buri 102
Chosun Ilbo 90
Chrysler 221
CIDA 69
cities 8, 13, 78
citizens 6, 9-10, 20, 25, 27, 42, 78, 81-82, 88-91, 145, 171, 206, 272, 274
civic structures 272
climate 49
co-curricular activities 165

CO_2 see carbon dioxide
coal 12
coastal environments 5, 31-33, 49-52, 58, 66-67, 70, 101-03, 111, 117, 132, 137, 181, 184, 187
cockle 185
coconut plantations 102
coliform bacteria 104
collaborative observation 255
commune system 11
communications 201, 206, 211, 236
community 43, 183, 190, 224; see also *citizens*
computer 234, 256, 261-262, 275, modeling 211; technologies 225; and information 211-21
conflicts of interest 23-24
conservation 9, 21, 29, 42, 44, 71, 118, 144, 185
Consortium for Earth Science Information Network 219
construction industry 17
consumers 15, 26
consumption 189
contamination 211
COP3 199
copper 221
copyright 216
coral reefs 51, 97, 117, 185
corporate sector 44
corruption 6, 69
Crassostrea gigas 67
cultural minorities 173
curriculum 145, 164, 168, 171-72, 179, 230, 233, 236, 265
Curriculum Council 232

-D-

dam construction 4, 23, 56-59, 70
data 172, 214, 219, 224
decentralization 120
DECS 171, 177
deforestation 56, 62, 70, 96-97, 116, 132, 170, 184, 235, 2421
deltas 52, 57, 61
democracy 4, 80, 91, 128
demography 22; see also *population*
Deng Xiaoping 13
DENR 171
Department for International Development 69

Department of Agriculture 44
Department of Environment 32, 40-41, 44
Department of Environment and Natural Resources 171
Department of Fisheries 44
Department of Forestry 44
Department of Geological Resources 101
Department of Land Development 99
Department of Wildlife 44
dependent thinking 206
desertification 11, 61-62, 66, 70, 132-33, 202
detergents 20
development 4-6, 13, 24, 88, 114-15, 117; see also *planning, sustainability*
diffusion 205
dioxin 10
Director-General of Environment 35
disasters 19, 132-33
dissolved oxygen (DO) 102,104-05
Doe, L. 218
Dong-A Ilbo 90
Dongting, Lake 137
Doosan Electronics 82
droughts 19-20
dry-season cropping 101
drylands 63
Du Pont 9
dumping 10, 34

-E-

Earth Summit II 49
earthquakes 51, 70
ecology, see environment
economic growth, 2, 13
ecotourism 72
editorials 90
Educational Reform Program 234
EENET 226, 247-49
EIA 37-38
El Niño 6, 243
electricity 10, 13
electro-conductivity 150
electronic library resources 214
elementary schools 144-45, 147, 156, 237
endangered species 54, 99; see also *animals, fauna*
Endau-Rompin Reserve 44

energy 12-13, 70, 77, 79, 104, 106, 112, 116-17, 245
engineering 7
ENRA 122
enterprise web 225, 244, 276
entrepreneurship 30
environment 113, 146-47;
environmental accounting 122;
agenda 24, 242; awareness 26;
campaigns 81; data 224; degradation 25; diorama 177; global, 172;
groups 17, 80; housekeeping 205;
impact assessment 16, 37-38;
information 204; literacy 207;
movement 13, 16, 80; protection 114; resources 25; restoration 23, 25; standards 80; studies 233; see also *education, environmental*
Environment Agency 144
Environmental Education Network (EENET) 225
Environmental News Network 217
Environmental Organization Web Directory 217
Environmental Protection Agency 10
Environmental Quality Act 35
Environmental Quality Council 35
Environmental Sciences 124
EPA Toxics 219
EQA 35
erosion 52
estuaries 52, 57; see also *deltas*
ethics 199
Eutrophication Control Ordinance 20
eutrophication 7, 20, 22, 79, 137,139,151,203
evaporation 63
examinations 230
experiential learning 158
extension education 253

-F-

Faculty of Education, Shiga University 261
falaj system 52
family businesses 7
farmers 54, 81, 100,115, 128, 211; see also *agriculture*
fauna 27, 31, 45, 70, 97; see also *animals*
fertilizers 11, 28, 79, 166
fires 244

First Chemical Market 221
fish 2, 7,9, 31, 39-40, 57-58, 81, 98, 102-0,139,170,185,187
Fisheries Department 1896-87
flexibility 268,275-76
flexible/distance learning 226
flies 8
floods 4, 19-20, 31, 52, 57-58, 64, 70, 100-01, 181
flora 27, 45, 70, 97; see also *deforestation, forests, plants*
Ford Motor Company 221
Forest Love Water Project 182
Forest Research Institute 44
forests 4,5, 12, 21,24,40,44,47,55,81,101-02,111,115,117,132,182-84,201,224,272; conservation 128; depletion 4; fires 243-44; lands 22; resources 190; see also *deforestation*
fossil fuels 12
fuel 12, 32, 1042 see also *energy*
Fujian 15
full text 214-16
fund raising 128
future 122,127-28

-G-

game reserves 55
Ganges 51, 57, 68
garbage 8, 151; see also *wastes*
gasoline 123
GDP 29-30,93,94,112,190-92,196,
General Motors 221
GIF 213
Gilgit 56
glaciers 55
glass 8
Global Recycling Network 220
global warming 24,49,68,132,143-44,235,242,273
golf 18, 39, 81-82
government, agencies 224, budget 118, organizations 127, policies 117, role of 95,128,187
graphics 213
grazing 133
Great Lakes Information Network 219
Great Leap Forward 13
green agricultural products 128
greenhouse effect 68-69, 189, 201
ground water 58, 66, 101, 211

groupers 186
Guangdong 15
Gulf of Thailand 181

-H-

Hainan 12
handicapped 155
handicrafts 128
hazardous wastes 7-8, 24,32-33, 35, 39, 109, 112, 117, 120-21, 189, 201
haze 5-6, 32,251
health 7-8,22, 28, 67, 72, 81,109, 111,121,125-26,139,144,153,170, 189,193,213,221,227
heavy metals 8, 103, 109-10
hepatitis 8
heritage sites 49,65,72
Hiarashi 256, 258-59
high schools 144,147-48,154, 156, 230-31
high-order thinking 169,170,175-79
higher education 169,171, 179, 234, 236, 239, 245-47, 251,,269; see also *universities*
Himalayas 52, 54, 65
Hindu Kush 54-55,73
Hong Kong 2, 15, 93
Hotaru-Das 256
hotels 17, 101
Houchin 9
housewives 20
housing 12, 15-16, 34, 85,101
Hua Mark 101,107-08
Huk Muang Nan 184
human resources 26, 109, 113,117, 120, 194, 197, 231; see also *education*
Hunza 55
hurricanes 50, 51
hydrocarbons 8, 72
hydrochlorofluorocarbons 68
hydroelectricity 58, 106
hypertext markup language (HTML) 213
hypertext-based technology 216

-I-

IBM 216,218
Ikiruchikara 229
Ilan 9

Index

Inchon International Airport 91
incineration 4, 8. 78
income distribution 85,113,117
India 51
Indonesia 1-2, 4-6, 11,15-16, 18,43, 51, 60, 93, 189-93,196, 243-44, 276
Indonesian Agenda 21 193
Indus 49-53, 57-67, 74-76, 273
informal education 5, 124, 169-71, 181-82
Information Processing Center 261
information age 212,250; literacy 231; systems 70-76; technology 230-31; see also *education, computers*
Infoseek 214
infrastructure 3,16,18,20,103,112-13,226,235,246-47,261
Inland Sea 2, 203
input-output analysis 205
Institute for Science and Mathematics Education Development 169-70, 176
Institute for the Promotion of Science and Technology 164
Institute for Water Research 211
integrated studies 157, 160, 233
interdisciplinary teaching 164
International Environmental Education Workshop 143, 271
International Lake Environment Committee (ILEC) 131, 137, 140-41, 156, 166, 185, 221, 274; Environmental Education Project 140-41
Internet 204, 212-13, 215, 217-19, 222, 226, 233, 235, 246, 263-64
Internet Explorer 213
investment 31
Ipoh 32
irrigation 11, 27, 34,39, 52, 57-58, 62-63, 66, 114, 181-83, 243, 273
ISI Citation Databases 216
Islamabad 56, 58, 72, 76
issues 200-209
Itai-itai disease 3, 144

-J-

Jakarta 17,189,196,197
Japan Inc. 16
Japan Meteorological Agency 256-57
job bank 218
Johor Bahru 32

Jomtong District 183
Joto Elementary School 235
journals 9, 73, 166-167, 204, 207, 209, 215-216, 249, 277
JPEG 213
junior high schools 144, 147-148, 154, 156, 237; see also *education, environmental*

-K-

Kaka 17, 67
Kalimantan 244
Kansai airport 17
Kaohsiung 9
Karachi 58, 67-68, 72-75
Karakoram 51, 54-55
Kepong 44
Kho Samed 102
Kho Samui 105
King, of Thailand 182
Kita-Daito Elementary School 235
KNOT Plan 234
Kobe 17, 139
Korea 1, 2, 4, 7, 9, 15, 16, 77-82, 86, 88-93, 243, 273
Krabi 103, 187
Kuala Lumpur 32, 44, 46-47
Kuala Selangor 44
Kuala Selangor Nature Park 44
Kuching 32
Kumamoto 3
Kusatsu 131, 141, 166, 253
Kyoto 19, 49, 139, 199

-L-

Ladprao 101
Lahore 58, 73-74, 76
laissez faire 6, 123
Lake Biwa, see Biwa, Lake
Lake Biwa Comprehensive Development Project (LBCDP) 19-21, 27, 154
Lake Biwa Day 158
Lake Biwa Museum 253, 255, 257, 275
Lake Biwa Ordinance 204
Lake Biwa Research Institute 19, 154
Lake Environment Training Center of Shiga University 157

lakes, 23, 27-28, 44, 131, 137-139, 141, 149-150, 152, 154, 166, 219, 257; see also *Biwa, Lake*
Lampang Province 104, 123
land 1-2, 4-5, 8, 11, 22, 24, 28, 31, 33-34, 39-40, 49-50, 52, 54, 56-58, 61-62, 64-65, 67, 70, 73, 81, 96-101, 109, 111, 114-115, 121, 130, 132, 134, 138, 177, 181, 183-184, 187, 205, 223, 226, 242-243
reclamation 2, 4, 67
landfills 8, 10, 33, 78, 91
landslides 55, 70
lead 40, 62, 107-108
Leadership for Environment and Development (LEAD) 221
legislation 3-4, 9, 12, 29, 33-35, 37, 40, 42, 102, 219, 247, 273-274
leisure 15-18; see also *tourism*
Liaison Committee on Science Education 236
libraries 215-216, 247, 249
Life Studies 144
life expectancy 13
Lifelong Learning 145-146
lignite 104, 106
loans 69
local area network 261
local government 3, 19, 34, 43, 121, 177
local people 3, 12, 117, 133, 185, 187, 253, 255
logging 4-6, 31, 50, 61, 63-65, 69-70, 97, 114
Love River 9
Lukang 9
Lumniztera recemosa 67

- M -

Mae Klong 102, 104-105
Mae Moh 104, 123
Maeda Elementary School 235
Mahidol University 124-125, 130
Malayan Nature Society 44-45, 47
Malaysia 1-2, 4, 6, 15-16, 18, 29-31, 33-35, 39-41, 43-47, 49, 66, 93, 243, 273, 276
Malaysian Nature Society 276
management 19, 24, 28, 33, 35, 181, 183, 186-187
Mangla Dam 58-59

mangroves 31, 44, 51-52, 58-59, 66-68, 75, 97, 101-102, 117, 151, 181, 184-187
Manila 129, 170, 179, 196
manufacturing 27, 29-30, 33, 90, 112, 219, 220, 273; see also *industry*
manure 79; see also *fertilizers*
marine biotechnology 67
marine environments 2, 32-33, 37, 67, 185; see also *aquatic* environments, coral reefs, fish, shrimps
market economy 133, 273
market failure 113, 123
Material Safety 219
materials cycle 174 see also *recycling, wastes*
McKean, M.A. 3, 16
media 4, 11, 29, 35, 42, 72, 80, 82, 85-86, 89-90, 95, 176, 181-183, 216-217, 224-225, 246-247, 250, 261, 263, 275
media tools 261-262, 266-267
medicinal herbs 184
Mekong 51
mercury poisoning 3, 214
mesoplankton 186
metals 8, 103
meteorological data 224, 256, 259
methane 68-69
Metropolitan Water Works Authority 101
Michigan State University 211, 214, 216, 219
middle class 15-16, 127
migration 112
Minamata disease 3, 144, 214
Mineral Development Act 32
minerals 6, 71, 190; see also *mining*
mining 4, 31, 33-34, 40, 101, 104, 114
Ministry of Education 125, 144, 148, 232-235
Ministry of International Trade and Industry 235
Model Schools 156-157, 274
Moghul period 52
Mohenjo Daro 65
monitoring 4-7, 15, 38-39, 50, 70, 81, 107, 116, 120, 125, 157, 170, 177, 195, 243, 245, 250
monks 182, 224
monoculture 181
monsoon 52, 55, 57

motivation 89, 143, 151, 230, 267, 276
motor vehicles 4, 32
motorcycles 10
mountains 2, 49-53, 55, 57, 70, 86, 133, 235, 273
MSU Extension Service 211
muang fai 183
mud crabs 186
Murree 56
Museums 231, 253-255

-N-

Nakdong River 78, 82
Nan Province 184
National Center for University Entrance Examinations 230
National Economic and Social Development Board 93
National Economic and Social Development Plan 113
National Environmental Education Action Plan 169, 171, 275
National Environmental Training Center for Small Communities 220
National Forestry Council 40
National Land Council 40
National Mineral Policy 32
National Pollution Prevention Center 220
national parks 5, 8, 97
Natural Resource and Environmental Accounting 121
Nepal 54
netizen 206
Netscape 213
networks 201, 223-226, 233-234, 244, 246, 250, 262, 266, 268, 274, 277
New Omi Culture 154
New York 49
New York Times 216
Newly Industrializing Country 30
news groups 218
non-governmental organizations (NGOs) 1, 5, 9, 26, 29, 42-46, 71, 80, 82, 86, 88, 95, 113, 116-117, 121, 124, 126-128, 164, 166, 182-187, 250, 272-276
Niigata 3
NIMBY 109
nitrogen 8, 46, 150, 153
nitrogen oxide 243

nitrous oxide 68-69
noise 8, 36, 104, 109, 111, 129, 242
non-formal Education 125, 163
North West Frontier Province 64
Northern Areas (Pakistan) 55, 64
nuclear energy 10, 90, 202
nuclear war 83
nuclear waste 81

-O-

ocean pollution 143
ODA 69
Office of the Environment 77
Office of the National Environment Board 186
Office of Water Programs 220
oil spills 33
Okinawa 235
Omi 19, 154
Onsan 81
Osaka 17, 19, 139, 207
Otsu 261-262
over-harvesting 98, 273
Overholt 2, 13, 15
oxides 8
oxygen 28, 46, 102
oysters 67
ozone 132, 143, 189, 201, 242

-P-

Pacific Ocean 243; see also *Asian-Pacific*
Pakistan 51, 55-56, 58, 61-68
palm oil 35-36, 40
Papua New Guinea 66
participation 3, 25, 38, 68, 71, 81, 91, 113, 115-116, 120, 123-128, 184, 187, 190-191, 194, 200, 217, 260, 272, 275; see also *citizens*
participatory learning 159, 275
particulate 32, 57
Pasak 104
pasture 54, 133; see also *agriculture*
Pattani 151, 166
Pattaya 102
peer review 214-216
Pergau dam 31
pesticides 8, 28, 103, 109, 111; see also *chemicals*
Petchburi 185
Phangna 187

phenol 78, 82
Philippine Environmental Code 171
Philippines 51, 66, 93, 169-173, 177-179, 207, 276
phosphorus 20, 150, 153
Phrakhanong 101
Phuket 187
phytoplankton 27, 153
pine forest 183
planning 1, 4, 6, 12, 22, 31, 35, 37-40, 76, 114-115, 120-122, 128, 165, 187, 190, 200-203, 207, 255, 257, 261, 273, 276; Biwa, 19-28; Malaysia, 29-47; Thailand, 113-117, 121-122
plankton 21, 186, 203
plants 2, 31, 51, 54, 69, 78, 98, 104, 117, 123-124, 166, 173, 190, 202, 219, 237
plastics 7, 221
plug-ins 213
poaching 97-98
police 111
policy, see planning
politics 3-4, 16; see also *citizens*
pollutants 4, 22, 32, 58, 78, 103-104, 109, 174, 243-245, 273
polluter-pays principle 117
pollution 2-12, 20-21, 26, 31-37, 40, 49, 66, 77-78, 90-92, 98, 103-104, 112, 116-126, 132, 137, 143-144, 151-153, 181, 189, 201-206, 211, 220, 242-243, 272-276; see also *air, wastes, water*
polyaromatic hydrocarbons 8
population 7-16, 27, 49, 52-54, 57-59, 66, 70-72, 78, 86, 95, 98-99, 102, 112, 132-134, 139, 143, 158, 171, 174, 185, 189, 194, 201, 242, 271
Portney, P. 206
potential evapotranspiration 62
poultry 56
poverty 72, 100, 113, 126, 171, 189, 191-193, 272
poverty alleviation 72
power stations 4, 32, 243
practical experiences 174, 238
prawns 32; see also *shrimps*
precipitation, see rain
Prefectural Board of Education 150, 153-157, 274
Prince of Songkla Universities 151

private sector 29, 42, 44, 46, 116, 120-123, 176, 276
privatization 30, 117, 120
problem-solving 159, 249
property rights 116
protests 3-4, 9, 13, 17, 78, 109, 272; see also *citizens*
Public Awareness 77, 92
public 26-27, 29, 40-46, 71, 78, 90-91, 95, 125-128; see also *citizens*
Punjab 58, 64-65
Pusan City 81
push-net boats 181, 186

-Q-

Queen, of Thailand 182

-R-

radioactive wastes 8
rain 8, 12, 31-32, 55, 133
rationality 206
rats 8
reclamation 4, 80
recycling 8, 91, 199, 220-221
red tide 20, 28, 153, 203
reforestation 114
regulations 36-37
Religion for Society Foundation 182
reservoirs 28
residents 23; see also *citizens, public*
resort development 15, 31, 80; see also *golf, tourism*
Resources for the Future Foundation 206
retailing 17
Revelle Report 65
Rhizophora mucronata 67
Rho Tae Woo 80
rice cultivation 183
Rio de Janeiro, conferences 24, 49, 193, 242
River Watch 177, 276
rivers 7, 31-32, 49-52, 56-58, 65, 70, 102-104, 137-138, 140, 143-144, 151, 189, 201, 206, 237
roads 13
Royal Forestry Department 183
rubber 29, 36
rust 86

Index

-S-

salination, see salt
salinity, see salt
salinization, see salt
salt 5, 57-58, 70-71, 100, 138, 244
salt water 8
Samut Sakhon 102
*Samut Songkram 102,185
Sarawak 31
Science and Mathematics Education Manpower Development Project 177
Science and Technology Basic Plan 234, 237
Science Council of Japan 236
scientists 253
sea level 68, 70, 201
seagrass 184, 188
search engines 212
seaweed 67
sediment 75
seismic activity 51
Selangor River 44
Seoul 78
Seoul-Pusan Bullet Train 91
Seta River 27, 133
Seto Inland Sea 203
sewage 32, 36, 39, 102-103, 139
Shanghai 15
shellfish 98
shifting cultivation 4, 31, 183
Shiga Prefecture 19-20, 28, 131, 143, 155-156, 199-200, 203, 253, 261, 274-276
Shiga Project for Environmental Education 150
Shiga University 152, 209, 226, 261, 267-268
Shikoku 2
Shinsul-dong 78
shipping 2
shrimp 184
Siam Environment Club 182
Sichuan 12
Sindh 58, 62, 64-65, 67
Sindh Forest Department 67
Singapore 2, 6, 15, 93, 243
skills 172-175
slash-and-burn cultivation 183
Small House in the Dense Forest Project 182
smog 4, 78
smoke 5

Social Studies 144
social movements 13
sodification 61, 64
soil 39, 50, 55, 64-65, 67, 98; erosion 5, 8, 11, 40, 50, 98, 135, 137, 242; exhaustion 101
solid wastes 4, 22, 32-33, 78-79, 90, 273; see also *hazardous wastes, sewage, wastes*
South 207
South China Sea 33
South Korea, see Korea
Southeast Asia 24, 243; see also *Indonesia, Malaysia, Thailand*
Southern Artisanal Fisherman Federation 186
Space Collaboration System 236
sports 171
spreadsheet 264
state enterprises 13
steel 221
stewardship 174
Stockholm, conference 143, 190, 272
STORET environmental data systems 214
Straits of Melaka 33
structural adjustments 29
students 165, 220, 263, 265, 267; see also *universities*
subsidence 8, 101
Suharto regime 6
Suita City 207
Sukhumvit 101
sulfur 8
sulfur dioxide 78-79
Sultanate of Oman 52
sustainability 47, 192, 196, 244
Synthetic Organic Chemicals Manufacturing Association 219

-T-

Taipei 8, 10
Taiwan 1, 2, 6-11, 15, 16
Tamsui river 10
Tanaka Kakuei 17
Tanakami Mountains 136
Tanganyika 27
Tarbela Dam 56, 58-59
Tbilisi 271
teaching 169, 171, 179, 236, 276; materials 143-152; Philippines 169-179; Shiga 153-161; Thailand 163-

166; with computers 261-68; see also *education, environmental*
tectonic activity 51
temperature 54
Texas Clean Air Act 219
text 213
textbooks 155
Tha Chin 104, 117
Thailand 1, 2, 4-6, 15-16, 93-106, 120-131, 151, 163-166, 185-188, 197, 224, 276
Thammasat University 125, 130
Three Gorges Dam 13
Tianjin 11, 15
timber 114; see also *forests, deforestation*
tin 4, 29
Tokyo 17
tourism 31, 71-72, 94-95, 102, 116, 187
toxic wastes 189; see also *hazardous wastes*
Toyama 3
traffic 104
Training and Development Resource Center 218
training 157; see also *education, environmental; teaching*
Trang 184-186
transport 12, 17-18, 52, 113
trawlers 181, 184, 186-187
trees 4, 12, 56, 63, 124, 182, 184, 224; see also *forests, deforestation*
Tsutsumi family 17
turbidity 150
Tyres Report 65

-U-

Ulsan 81
Uminoko 154, 157, 274
UNCED 199
UNDP 54, 67
UNEP International Environmental Technology Center 219
UNESCO 69, 207
United Nations 62, 143, 245
United Nations Conference on the Human Environment 143
United States Department of Commerce National Oceanic and Atmospheric Administration 219
United States Department of Energy 219
United States Environmental Protection Agency (USEPA) 219
United State government 211
United State National Information Infrastructure Initiative 235
universities 43, 45, 72, 116, 124-125, 151, 176, 195, 201, 204, 223-225, 230, 232, 236, 246-247, 250, 259, 261, 277
UP-ISMED 170, 176-177
Urazoe City 235
urban areas 4, 32-33, 103, 111-112, 116, 125, 132
Uroglena americana 27
USSR 138

-V-

values 25-26, 43, 72, 104, 159, 161, 175, 192, 202, 253, 275
vehicles 4, 10, 32, 221
Viboon Kemchalerm 184
video 171, 213, 216-217, 236
video conferencing 217
Vietnam 2, 49, 243
virtual reality 211, 213
VRML (virtual reality modeling language) 213

-W-

Wakayama 17
WAPDA 58, 61, 65
waste water 4, 21, 23, 78, 80, 103-104, 112, 137, 153, 219, 220
wastes 4, 7-8, 22, 26, 32-33, 78-79, 90-91, 102-103, 112, 170, 173, 189, 201, 273; see also *hazardous wastes, solid wastes, waste water*
Wat Chan 183
Water and Power Development Authority 58, 65
Water Quality Index 32, 46
waterlogging 61, 63-65, 70
watershed 19, 21-22, 28, 96, 117, 131, 137, 140-141, 150-151, 183
weather patterns 68
web sites 212, 216-221, 225, 247, 249-250, 277
West Virginia University, Morgantown 220

wetlands 4, 31, 58, 80; see also *coastal environments, mangroves*
Wildlife Fund of Thailand 182, 186
wildlife 8-9, 40, 55, 97, 161; see also *animals, fauna*
wind 79, 256-59, 275; erosion 64, 133; observation 256
wood pulp 6
word processing 217, 263-265
World Bank 69, 122, 242, 245, 250
World Commission on Environment and Development (WCED) 190, 272
World Conservation Strategy (IUCN) 190, 272
World Heritage Sites 65
World Wide Web (WWW) 212-213, 216-220, 247, 263, 277
WWW Chemicals 221

-Y-

Yad Fon 184, 186
Yahoo 213, 219
Yangtze River 51, 134, 137
Yellow Sea 79
Yeochun 81
Yodo River 19
Yuki-Das 256-257
Yunnan 134-135
yutori 229

-Z-

zest for living 229, 230, 275, 277